Mustard Lung

Mustard Lung

Diagnosis and Treatment of Respiratory Disorders in Sulfur Mustard-Injured Patients

Mostafa Ghanei
Baqiyatallah University of
Medical Sciences, Tehran, Iran

Ali Amini Harandi
Shahid Beheshti University of
Medical Sciences, Tehran, Iran

AMSTERDAM • BOSTON • HEIDELBERG • LONDON
NEW YORK • OXFORD • PARIS • SAN DIEGO
SAN FRANCISCO • SINGAPORE • SYDNEY • TOKYO

Academic Press is an imprint of Elsevier

Academic Press is an imprint of Elsevier
125 London Wall, London EC2Y 5AS, UK
525 B Street, Suite 1800, San Diego, CA 92101-4495, USA
50 Hampshire Street, 5th Floor, Cambridge, MA 02139, USA
The Boulevard, Langford Lane, Kidlington, Oxford OX5 1GB, UK

British Library Cataloguing-in-Publication Data
A catalogue record for this book is available from the British Library

Library of Congress Cataloging-in-Publication Data
A catalog record for this book is available from the Library of Congress

ISBN: 978-0-12-803952-6

For information on all Academic Press publications visit our website at https://www.elsevier.com/

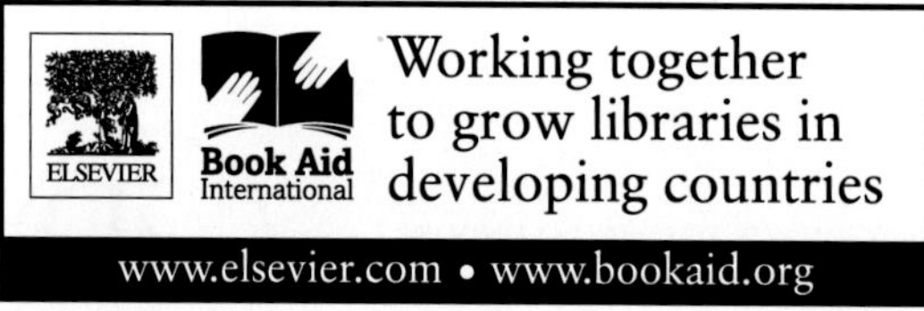

Typeset by TNQ Books and Journals
www.tnq.co.in

Contents

Preface

The use of chemical agents in wars as an effective weapon to disable or eliminate the enemy has a long history. During the Iraq–Iran war of 1980–1988, more than 200 chemical attacks were mounted against the Iranian armed forces and civilians. As a result, more than 100,000 noncivilians and civilians are currently suffering the late complications of chemical injuries. Considering the large number of chemical warfare victims in Iran and the fact that the first chemical attack by Iraq against Iran was carried out more than two decades ago, it seems necessary to perform and conduct comprehensive research on the late complications of these agents.

Considering the history of the use of mustard gas (sulfur mustard) in wars and its production in industrial factories, the conducted investigations on the early and late complications of mustard gas can be divided into studies performed on the casualties of World War I, workers of the production factories, and the victims of the Iraq–Iran war. It should be noted that most investigations have focused on the first two groups. Moreover, they have mostly addressed the early and late complications of mustard gas. However, since these studies are old, they lack the modern and advanced diagnostic and therapeutic tools and equipment. Moreover, methodologic defects such as disregarding the confounders and lack of an appropriately selected study group are other limitations of these studies. Investigations performed in other countries have had less access to the studies performed in Iran. Hence, their investigations on the late complications of these agents are brief and incomplete. For this reason, not only can the results of these studies not be used by Iranian researchers but also they sometimes fail to meet the demands of the physicians and patients for treatment and control of the complications of the chemical injuries.

The conducted investigations on the late complications of mustard gas have been mostly presented as dissertations, congress papers, and research projects published in scientific journals and are therefore disorganized and sometimes incomplete. Furthermore, often research carried out in domestic scientific centers is not purposeful and has not been performed based on the diagnostic and therapeutic priorities to relieve the problems of the affected people.

In this book, we have reviewed and discussed the history, classification of chemical warfare, mechanisms of injury, related respiratory disorders, preventive measures and approaches, physiopathology, biomarkers, diagnostic methods, gastroesophageal reflux, mustard gas carcinogenesis, and new treatments. Furthermore, prescriptions for the chronic phase have been included as an appendix to this book.

I hope that this book, which has been written following a thorough review of the credible international and domestic literature and based on my experience of years of clinical practice and research is useful and practical for fellow physicians and medical students in order to take effective steps to help the chemical warfare victims.

Finally, I would like to thank Dr. Hamidreza Ghadimi, MD for translation of the book from Persian to English and all who helped me with writing and publishing this book.

Professor Mostafa Ghanei
Internist, Pulmonologist
Baqiyatallah University of
Medical Sciences, Tehran, Iran

CHAPTER

History of Chemical Weapons Use

1

The use of chemical agents as a weapon has a long history. Chinese warriors were familiar with arsenic gas about 1000 BCE [1]. About 800 years after that, Sun Tzu, a Chinese general, wrote of a firing weapon in the book *The Art of War*. Moreover, in 200 BCE, it was explained the use of bellows to pump smoke from burning balls of mustard and other toxic vegetables into tunnels being dug by a besieging army. The first use of chemical weapons in the West goes back to the Peloponnesian War in 500 BCE. They produced toxic fumes by burning wood soaked in sulfur.

The repeated use of chemical agents in human wars was the most inhumane war tactic of the past century. Therefore, researchers in different fields of science, especially medical researchers, have made efforts to find appropriate preventive and therapeutic measures.

Chemical weapons were first officially used in World War I (WWI, 1914–1918). In 1894, the British forces had made ineffective use of sulfur gas against Russian troops in the Crimean War. However, it was the Germans who used chlorine effectively against Belgian soldiers stationed in Ypres on April 27, 1915 [2]. This attack was the first massive gas attack in war history, declaring that new warfare entered the battlefields. Since the Belgians were taken by surprise, were not prepared, and lacked protective gear such as masks, the attack resulted in the death of 5000 and injury of 15,000 Belgian soldiers and other troops in the region. Hence, efforts were started to both produce and confront chemical agents [3]. The use of chemical warfare in WWI and other wars became an increasing trend and resulted in many casualties. It is interesting that chemical weapons were not extensively used after World War II (WWII). After that war ended, extensive research was started on chemical agents, making the production of more potent chemical agents possible. Mustard gas was possibly developed in 1822 by César-Mansuète Despretz although he never mentioned its blister-forming properties when describing the reaction of sulfur dichloride and ethylene. In 1854 a French chemist named Riche reported a similar reaction but did not describe any harmful physiological properties. In 1860 Frederick Guthrie, a British scientist, synthesized mustard gas, described its chemical characteristics, and made mention of its unique smell and irritating properties [4]. In the same year, a German chemist named Albert Friedrich Emil Niemann, who was a pioneer in producing cocaine, described the blister-forming properties of mustard gas. Moreover, Viktor Meyer published a paper describing the process of combining 2-chloroethanol with aqueous potassium sulfide, resulting in the

Mustard Lung. http://dx.doi.org/10.1016/B978-0-12-803952-6.00001-0

production of thiodiglycol with phosphorus trichloride. The produced compound was much more potent and its adverse effects were more severe. He first noticed the adverse effects in his assistant, which he thought were psychosomatic symptoms, but when he tested the compound on laboratory mice, most of them died. In 1913 Hans Thacher Clarke, a British chemist, replaced phosphorus trichloride with hydrochloric acid in the Meyer's formulation when he worked with Hermann Emil Louis Fischer. He was hospitalized for 2 months for burns when the glass containing the produced compound broke. The report of the society of chemists on this incident to the German Empire paved the way for the use of this new agent as a chemical weapon [5]. In WWI, the German Empire developed deadly chemical weapons with the 2-chloroethanol chemical structure using the Meyer-Clarke method. In fact, mustard gas or sulfur mustard was the first vesicant agent used as a chemical weapon. Mustard gas caused a persistent contamination in the environment, and chemical war entered a new era with its introduction to the battlefields.

In July 1917 the German Army used sulfur mustard, which had a distinctive garlic-like odor, against the Allies, with devastating results. In total, the use of the chemical weapons resulted in the deaths of about 100,000 and injury of 120,000 people during WWI [6] (Table 1.1).

Although countries had large stockpiles of chemical weapons, they did not make much use of them in WWII. When the war ended, many countries disposed of their chemical weapons in different ways. Most of the sulfur mustard gas in Germany was dumped into the Baltic Sea. Between 1966 and 2002, about 700 chemical weapons, mostly sulfur mustard, were found near Bornholm, an island near Denmark. Mustard gas depots are still found in different parts of the world. After WWII, sulfur mustard was disposed of by explosion undersea, which was discontinued because environmental regulations prohibited it, and this method was replaced with other means such as disposal in factories. Unfortunately, the chemical agents left from WWI, despite their age, are still very potent and lethal. Some chemical weapons dumping sites under the sea have been lost in the course of time or only their rough locations are known due to poor recordkeeping.

Table 1.1 Some Examples of the Use of Mustard Gas as a Chemical Weapon

World War II started about 20 years after World War I ended. WWII is important because during this time, biological and chemical weapons, including mustard gas, were developed like never before. This table lists some examples of the use of mustard gas as a chemical weapon from the first use to date.

- United Kingdom against the Red Army in 1919
- Spain and France against the rebels in Morocco during 1921–1927
- Italy in Libya in 1930
- The Soviet Union against China in 1934
- Italy against Ethiopia from 1935 to 1940
- Japan against China during 1937–1945
- Egypt against North Yemen during 1963–1967
- Iraq against Iran during 1983–1988
- Sudan against insurgents in the Civil War, in 1995 and 1997

Numerous instances of accidental exposure to mustard gas have been also reported, both in the staff responsible for its depot and incineration and the civilians who were accidentally contaminated. In 2010 some old artillery shells from WWI from the Atlantic Ocean near New York severely injured some fishermen. Unfortunately, these accidental exposures and contacts with low doses of mustard gas can result in severe complications and mortality in the long term, similar to a heavy chemical bombardment.

In 1960 talks started in Geneva to ban the use of chemical weapons, which paved the way for the 1972 Convention on the Prohibition of the Development, Production and Stockpiling of Bacteriological (Biological) and Toxin Weapons and on Their Destruction (BTWC). However, since this convention did not explicitly prohibit the "use" of biological weapons, the talks were resumed in 1981 and resulted in the Chemical Weapons Convention (CWC) in 1997.

The massive use of chemical weapons against Iranian armed forces and both Iranian and Iraqi civilians during the Iraq–Iran war is a very prominent example of the use of these agents in the recent decades [7]. In November 1980, Iraq mounted the first chemical attack against the Iranian city of Sousangerd. Nonetheless, the use of the chemical warfare was not recognized due to the lack of knowledge and preparedness to deal with these agents from the Iranian forces, and, therefore, no effective preventive measures were implemented. Unfortunately, the Iraqi forces continued to use chemical agents against Iran repeatedly during the war. The agents that were commonly used were nerve gas, mustard gas, cyanide, and sarin. Despite the massive use of chemical agents by Iraq, most of the victims were rescued using the classic treatment of nerve gas. However, the Iraqi troops even attacked the field hospitals several times. For example, a field hospital in Soumar was heavily contaminated by mustard gas, and a large number of medical staff working in the hospital were severely injured. In the winter of 1987, Iraq was officially introduced as the user of chemical agents against Iranian forces for the first time. The Iraqi Army mounted 252 chemical attacks against Iranian troops, logistic and supply centers, cities, and villages [8]. In the end, the total number of chemical victims was reported at more than 100,000 people, and more than 5000 people died due to complications from the chemical agents, of which about 1000 deaths were due to sulfur mustard and the rest due to the nerve agents and cyanide compounds. Currently, more than 50,000 chemical patients have medical records and are under treatment [9].

REFERENCES

[1] Richardt A. CBRN protection: managing the threat of chemical, biological, radioactive and nuclear weapons. Germany: Wiley-VCH Verlag & Co; 2013. p. 4.

[2] Papirmeister B, Feister AJ, Robinson SI, Ford RD. Medical defense against mustard gas: toxic mechanisms and pharmacological implications. United States (Florida): CRC Press; 1991.

[3] Ghabili K, Agutter PS, Ghanei M, Ansarin K, Panahi Y, Shoja MM. Sulfur mustard toxicity: history, chemistry, pharmacokinetics, and pharmacodynamics. Crit Rev Toxicol May 2011;41(5):384–403.

[4] Guthrie XIII F. On some derivatives from the olefines. Q J Chem Soc 1860;12(1):109–26.

[5] Duchovic RJ, Vilensky JA. Mustard gas: its pre-World War I history. J Chem Educ 2007;84(6):944.

[6] Veterans at risk: health effects of mustard gas and lewisite. Committee to Survey the Health Effects of Mustard Gas and Lewisite, National Academy of sciences, Institute of Medicine, National Academy Press, Washington, DC; 1993.

[7] Security Council of the United Nations. Report of specialists appointed by the secretary general to investigate allegations by the Islamic Republic of Iran concerning the use of chemical weapons. New York Security Council of United Nations; 1986. Document S/16433/1986.

[8] Afshar A. Chemical attack in Sardasht. ISNA; 2007. http://www.centralclubs.com/topic-t22503-12.html.

[9] Khateri S, Ghanei M, Keshavarz S, Soroush M, Haines D. Incidence of lung, eye, and skin lesions as late complications in 34,000 Iranians with wartime exposure to mustard agent. J Occup Environ Med November 2003;45(11):1136–43.

CHAPTER

2 Classification of Chemical Warfare Agents and Properties of Sulfur Mustard

DEFINITION AND CLASSIFICATION OF CHEMICAL WARFARE AGENTS

The general definition of a chemical weapon is a toxic substance that is contained in a delivery system like a bomb or shell. Although this definition is technically correct, it only covers a small proportion of the broad spectrum of warfare as "chemical weapons." For this reason, a more comprehensive definition is required that includes the munitions, chemical agents, and the equipment related to the production and use of chemical weapons. For example, in binary chemical weapons or munitions, a nonlethal chemical substance may be contained in a munition only to mix with a second chemical just before being fired, which produces a new chemical that is released when the target is reached.

According to the Chemical Weapons Convention, all toxic chemicals and their precursors that can cause death, temporary incapacitation, or permanent harm to humans or animals through their chemical action are known as chemical weapons [1] (except where intended for legal peaceful purposes in determined amounts like medical purposes). However, there are a number of unaddressed questions in the definition of chemical weapons like the role of domestic riot control agents. Moreover, whether dual-use chemicals are intended to be used as chemical weapons makes the definition even more complicated; for example, chemicals like chlorine, phosgene, and hydrogen cyanide, which were all used during World War I (WWI), are key intermediates in the production of numerous commercial goods. It should be mentioned that although the primary purpose of the research on chemical substances was never to reach their toxic properties, they turned to agents that could affect living beings via their skin or clothing. Gas masks are useless in these circumstances. For example, mustard gas can easily penetrate into the leather and fabric and inflict painful burns on the skin. During the 20th and 21st centuries, about 70 different chemicals were used or stockpiled for chemical warfare [2]. These agents may be in the form of a liquid, gas, or solid. The liquid agents that are easily evaporated are known as volatile or high-vapor-pressure materials. Many chemicals are made volatile so they can be dispersed rapidly over a vast area.

Chemical agents are classified as lethal and incapacitating. A substance is classified as incapacitating if less than 1/100 of the lethal dose causes incapacitation. There

Mustard Lung. http://dx.doi.org/10.1016/B978-0-12-803952-6.00002-2

is no fixed way to distinguish lethal and incapacitating substances, and their distinction relies on a statistical average known as LD_{50}.

Moreover, chemical warfare agents can be categorized according to their persistency defined by duration of time that a chemical agent remains effective after dissemination. Accordingly, chemical agents are classified as persistent or nonpersistent.

Nonpersistent chemical warfare lose effectiveness after a few hours or minutes or even after a few seconds. Purely gaseous agents like chlorine and highly volatile compounds like sarin are two examples of nonpersistent agents.

In addition to the agent that is used, the mode of delivery is also of crucial importance. To be deployed nonpersistently, the agent is disseminated as very small droplets that are comparable in size with the mist that an aerosol can sprays. In this form, not only the gaseous part of the compound (around 50%) is absorbed but also the fine droplets may be inhaled or absorbed through skin pores. An effective chemical weapon requires very high concentrations of the agent instantly; for example, one deep breath should contain a dose of the agent sufficient to kill. For this reason, the primary weapons that are used are rocket artillery, bombs, and large ballistic missiles with cluster warheads. With the use of nonpersistent agents, the contamination in the target area is low or undetectable after a few hours; for example, sarin or similar compounds cannot be detected anymore after about 4 h. On the other hand, persistent agents have a tendency to remain in the deployed area for several weeks to years, making decontamination complicated. Shielding is required for defense against persistent agents for long periods of time. Nonvolatile liquid agents, such as vesicants and the oily VX nerve agent, do not easily evaporate into a gas, and therefore remain hazardous for long periods of time.

The size of the droplets that are used for persistent delivery is up to 1 mm, which accelerates the falling speed; as a result, nearly 80% of the agent reaches the ground, causing heavy contamination. Therefore, the effects of the persistent deployment are not confined to the enemy.

Other likely targets of persistent agents include enemy flank positions (preventing possible counterattacks), regiments of artillery, commando posts, or logistic lines.

A special form of persistent agents is thickened agents, which consist of a common agent mixed with gelatinous thickeners to provide stickiness. Primary targets for this type of agent, considering their increased persistency and difficulty in decontaminating the affected area, include positions like airports.

The chemical agents are categorized into several major categories according to their mechanism of action. Table 2.1 presents their classification and characteristics [3,4].

PHYSICAL AND CHEMICAL PROPERTIES OF SULFUR MUSTARD

Mustard gas (dichlorodiethylsulfide or bis(beta-chloroethyl)sulfide (BCES)) was first made by Guthrie, a scientist who discovered its blister-forming characteristic, in 1859. Mustard is one of the earliest known alkylating agents. Boursnell et al. developed

Table 2.1 Classification and Characteristics of Chemical Warfare Agents

Class of Agent	Agent Name	Mechanism of Action	Signs and Symptoms	Rate of Action	Persistency
Nerve	Cyclosarin (GF) Sarin (GB) Soman (GD) Tabun (GA) VX VR Some insecticides Novichok agents	Inhibition of acetylcholinesterase in the synapses, causing both muscarinic and nicotinic effects	Miosis and pinpoint pupils Blurred vision Headache Nausea, vomiting, diarrhea Excessive secretions and sweating Muscle twitching and fasciculations Dyspnea Seizures Loss of consciousness	Vapors: seconds to minutes Skin: 2–18h	VX is persistent and poses a contact hazard; other agents are nonpersistent and cause damage mostly through inhalation
Asphyxiant/ Blood	Most arsines Cyanogen chloride Hydrogen cyanide	Arsine: Causes hemolysis leading to renal failure. Cyanogen chloride/hydrogen cyanide: Cyanide directly deprives cells of oxygen. The cells therefore use anaerobic respiration, resulting in lactic acid accumulation and metabolic acidosis.	Possible cherry-red skin Possible cyanosis Confusion Nausea Gasping for air Seizures before death Metabolic acidosis	Instant onset	These agents are nonpersistent and pose an inhalation hazard.
Vesicant/Blister	Sulfur mustard (HD, H) Nitrogen mustard (HN-1, HN-2, HN-3) Lewisite (L) Phosgene oxime (CX)	These alkalizing agents produce acidic components in the acute phase that damage the skin, eyes, and respiratory system. They also have destructive effects in the long term through oxidative stress.	Severe skin and eye pain and irritation Burns, blister, and skin infection Tearing, conjunctivitis, corneal damage Severe respiratory distress to severe airway damage	Mustards: vapors for 4–6h, eyes and lungs affected more rapidly while skin is affected after 2–48h Lewisite: immediate effect	These agents are persistent and pose a contact hazard.

Continued

Table 2.1 Classification and Characteristics of Chemical Warfare Agents—cont'd

Class of Agent	Agent Name	Mechanism of Action	Signs and Symptoms	Rate of Action	Persistency
Choking/ pulmonary	Chlorine Hydrogen chloride Nitrogen oxides Phosgene	Similar mechanism to blister agents with the production of acidic compounds in the acute phase causing respiratory damage and pulmonary edema and suffocation followed by severe pulmonary complications in the long term.	Airway irritation Eye and skin irritation Dyspnea and cough Sore throat Chest tightness Wheezing Bronchospasm	Immediate to 3h	These agents are nonpersistent and pose an inhalation hazard.
Lachrymatory agent	Tear gas Pepper spray	Cause severe irritation of the eye and temporary blurred vision	Severe irritation of the eyes	Immediate	These agents are nonpersistent and pose an inhalation hazard.
Incapacitating	Agent 15 (BZ)	Inhibits acetylcholine in the victim similar to atropine. Has peripheral nervous system effects opposite of those observed in poisoning with nerve agent.	Hallucinations, disorientation, and confusion Hyperthermia Ataxia Mydriasis Dry mouth and skin	Inhalation: 30min to 20h; Skin: Up to 36h (mean duration: 72–96h)	It is very persistent in soil and water and many surfaces and poses a contact hazard.
Cytotoxic proteins	Nonliving biological proteins like: Ricin Abrin	Inhibit protein synthesis	Incubation period of 4–8h, followed by flu-like signs and symptoms Progression within 18–24h Inhalation: Nausea, dyspnea, pulmonary edema Ingestion: Gastrointestinal hemorrhage and bloody diarrhea; sometimes liver and kidney failure.	Average: 4–24h. Inhalation or injection causes more severe signs and symptoms than ingestion.	These are mild agents that degrade quickly in the environment.

radioactive BCES to evaluate its metabolites and effects on different systems [5]. Then, Davison et al. produced radioactive mustard gas from radioactive hydrogen sulfide. Thiodiglycol (TDG) can be produced by reacting hydrogen sulfide with ethylene oxide. The reaction of thiodiglycol with hydrochloric acid yields mustard gas [3].

Later, Davison et al. used the modified Boursnell method for the synthesis of radioactive mustard in 1961 [6].

By keeping the ratio of the mass balance, hydrogen sulfide reacts with large quantities of ethylene oxide to produce TDG. The recent material is subsequently chlorinated with concentrated hydrochloric acid. Barium sulfide is used as the source of radioactive hydrogen sulfide.

Mustard gas has a molecular weight of 159.08 and consists of 30.20% C, 5.07% H, 44.58% Cl, and 20.16% S.

Mustard is a clear, colorless oily liquid with a boiling point of 215–217°C and a melting point of 14.4°C. In order to use mustard gas effectively at lower temperatures, it is mixed with Lewisite in some munitions at a ratio of 3:2. This compound has a melting point of −26°C. In its crude form, mustard is a pale yellow to dark brown liquid with a density of 1.278 g/mL. It has a distinctive onion- or garlic-like odor.

Sulfur mustard is poorly water soluble but its gaseous and liquid forms are readily soluble in oils, fats, and organic solvents and remain in compounds like alcohol, ether, benzene, and kerosene. Due to its high lipid solubility, it penetrates the lipid cell membrane very quickly. Sulfur mustard is a stable compound and is not affected by normal pressure and temperature. Its evaporation in air increases with increasing temperatures. It is slowly hydrolyzed in the water to thiodiglycol and hydrochloric acid (half-life: 5 min at 37°C). Both the liquid and gaseous forms of sulfur mustard readily penetrate ordinary clothing and fabrics. Table 2.2 shows the physical properties of mustard gas.

There are two types of mustard, including mustard gas and nitrogen mustard. They both have structural similarities and common primary chemical reactions. The key reaction of mustard is the intramolecular cyclization in a polar solvent (like water) producing a cyclic anion and free chloride anion. The cyclic form is responsible for the different effects of mustard.

In fact, mustards are a group of chemical compounds whose structure has a general formula of R-CH2CH2X where R is a leaving group and X is a Lewis base. The broad spectrum of the effects of mustards has made them useful in both medicine and chemical weapons. Table 2.3 summarizes different applications and chemical structures of mustards.

Three common mustard compounds are:

1. BCES (H)
2. β Ethyl-bis-(chloroethyl)-amine (E-BA)
3. β Tris-(chloroethyl)-amine (TBA)

These mustards are similar in all respects but nitrogen replaces sulfur in nitrogen mustards. Mustard evaporates very slowly under cold conditions while this substance is very dangerous at higher environmental temperatures (38–49°C) due its faster

Table 2.2 Physical Properties of Different Forms of Mustard Gas

Physical Appearance	Yellowish brown oily liquid
Odor	Similar to garlic, mustard, leek
Melting point	13.5–14.4°C
Boiling point	228°C
Density	1.27
Liquid density at 20°C	1.27 g/cm^3
Molecular weight	159.1
Density at 20°C	1.27 g/cm^3
Melting point	217°C
Boiling point	14°C
Evaporation at 25°C	900 mg/m^3
Water solubility at 25°C	0.068 g/L
Percutaneous LD_{50}	2.4 mg/kg (for male rats) 8.1 mg/kg (for female mice)
LCt50 (mg*min/m^3) for inhalation	1500
LCt50 (mg*min/m^3) for dermal contact	10,000
The smallest blister-causing dose on the skin	0.02 (mg)
Gas pressure	
At 10°C	0.032 mmHg
At 25°C	0.113 mmHg
At 40°C	0.346 mmHg

evaporation; for example, if the temperature increases from 10 to 38°C, the persistency of mustard in soil decreases from 100 to 7 h and the mustard in soil evaporates. For this reason, mustard gas was used at night in WWI to cause more damage and harm to the soldiers on the next sunny day. Therefore, the soldiers who thought they would not need protective clothes and masks the next day were severely harmed and injured. As a result, more than 80% of the mustard injuries in WWI were caused by mustard gas rather than liquid mustard. Since the density of mustard is 5.4 times greater than air, it tends to stay close to the ground and on the soil.

The term "mustard gas" is a common mistake. The compound that is known as "mustard gas" is a liquid that boils at 217°C. Liquid mustard and its gas are both vesicants.

In addition to mustard gas, many other mustard compounds have been examined and evaluated as potential chemical warfare agents. During WWII, some 100 types of mustard derivatives were produced, among which nitrogen mustards and sulfur mustard received attention. However, only mustard gas was extensively used during the war. Some nitrogen mustard compounds have peaceful applications and are used as mitosis inhibitors in the treatment of cancer. In WWI, the soldiers called it mustard for its distinctive odor. Some have compared its odor to the odor of garlic, mustard, horseradish, or leek.

Care must be taken not to mix mustard gas with mustard oil (allyl isothiocyanate), which is interestingly a blister agent.

Table 2.3 Applications and Chemical Structures of Mustard Compounds

Compound	Structure
Sulfur mustards	
Bis(2,2'chloroethyl)thioether (mustard gas)	$(Cl—-CH_2CH_2)_2S$
1,2-Bis(2-chloroethylthio)ethane	$Cl—CH_2CH_2—S—CH_2CH_2—S—CH_2CH_2—Cl$
1,3-Bis(2-chloroethylthio)-n-propane	$Cl—CH_2CH_2—S—CH_2CH_2CH_2—S—CH_2CH_2—Cl$
2-chloroethylchloromethylthioether	$Cl—CH_2CH_2SCH_2—Cl$
Bis(2-chloroethylthio)methane	$(Cl—CH_2CH_2S)_2—CH_2$
Bis(2-chloroethylthiomethyl)ether	$(Cl—CH_2CH_2SCH_2)_2O$
Bis(2-chloroethylthioethyl)ether	$(Cl—CH_2CH_2SCH_2CH_2)_2O$
Nitrogen mustards: military use	
Bis(2-chloroethyl)ethylamine	$(Cl—CH_2CH_2)_2NHCH_2CH_3$
Bis(2-chloroethyl)methylamine	$(Cl—CH_2CH_2)_2NHCH_3$
Tris(2-chloroethyl)amine	$(Cl—CH_2CH_2)_3N$
Nitrogen mustards: therapeutic use	
Cyclophosphamide	$(Cl\text{-}CH_2CH_2)_2N\text{-}P{=}O$, $(CH_2)_3—O$
Triscarboxymethylphenylbis(2-chloroethy) ethylamine	$HOOC(CH_2)_3\text{-}C_6H_4\text{-}N(CH_2CH_2\text{-}Cl)_2$
5-[Bis(2-chloroethyl) amino]-1H-pyrimidine-2,6-dione	$(Cl\text{-}CH_2CH_2)_2NH$, O, NH, O
N-3-bis(2-chloroethyl)-1,3,2-oxazaphosphinan-2-amide-2-oxide	O, P, O, $N\text{-}CH_2CH_2\text{-}Cl$ (H), $N—CH_2CH_2\text{-}Cl$
4-[Bis(chloroethyl)amino] phenylalanine	$(Cl\text{-}CH_2CH_2)_2NH\text{-}C_6H_4\text{-}CH_2CHCOOH$, NH_2

The symbol for mustard gas in military notes is H (abbreviated as HS), which may be taken from the word Hun-Stoffe. HN is also sometimes used for nitrogen mustard. During WWI, German forces used the acronym of LOST for mustard gas, which originated from the initial letters of the name of the German chemists who synthesized it (Wilhelm Lommel and Wilhelm Steinkopf). Since mustard gas was first used in Ypres, the French Army named it yperite. At that time, the Germans marked the artillery shells containing mustard gas with a yellow cross.

Mustard gas is 5.5 times heavier than air; therefore, the produced gas accumulates in the trenches and holes created by the blast. Mustard gas persists on the soil for a long time if there is no current of air or rainfall. Therefore, exposure to some amount of mustard even 2–3 days after the chemical attack is not unexpected. Hydrolysis of mustard gas is accelerated by increasing the temperature and pH. The water solubility of mustard gas is less than 1% while it is readily soluble in organic solvents like ethanol, ether, and chloroform. The fumes evaporated from the liquid mustard gas have a great penetrating ability and readily penetrate clothing, injuring the skin underneath.

Studies on the penetration of mustard gas have shown that about 80% of the liquid mustard gas that is poured on the skin evaporates. From the amount that penetrates the skin, only 10% remains in the skin, and the rest is absorbed systematically. Despite numerous investigations, the rate of the absorption of mustard gas through the respiratory system is unknown.

Although it has been stated that 64 mg/kg of mustard gas is lethal [7], the median lethal dose (LD_{50}) of mustard gas for percutaneous injection is not clear since the literature reports different amounts. The LD_{50} has been reported to be 9 mg/kg for percutaneous injection and 100 mg/m^3 for 10 min inhalation.

In female mice, the LD_{50} values were determined to be 5.7 mg/kg for percutaneous, 23.0 mg/kg for subcutaneous [8,9], and 8.1 mg/kg for oral routes [8], and 42.3 mg/m^3 for 1 h inhalation during a 14-day observation [10]. LD_{50} was also determined in male rats, which was found to be 2.4, 2.4, and 3.4 mg/kg through percutaneous, oral, and subcutaneous routes, respectively [8]. Prentiss [11] reported that contact with 0.15 mg/L of mustard gas for 10 min or with 0.07 mg/L for 30 min was lethal. There is a great emphasis on the toxicity of mustard gas. In general, contact with 0.07 mg/L of this compound for 30 min is considered lethal; therefore, it is five times more lethal than phosgene and 10 times more lethal than chlorine. Moreover, its physiologic half-life may also vary among species. The main routes of mustard gas absorption are through the lungs, skin, and eyes, along with the digestive system if swallowed. Approximately 80% of the sulfur mustard that reaches the skin is evaporated, 10% remains on the skin surface, and 10% is absorbed systematically. Mustard gas is distributed in the body in laboratory rabbits following intravenous injection and accumulates in the liver, kidneys, and lungs; about 20% of it is excreted within 12 h while the majority is excreted within 72 h. In dogs, sulfur mustard is distributed between the blood and tissues only within 5 min after inhalation [12]. The urinary metabolites of mustard gas include thiodiglycol and its conjugated compounds. Mustard gas is converted to other toxic substances like active threat agents at high temperatures; therefore, caution should be exercised when burning materials contaminated with sulfur mustard.

REFERENCES

[1] Convention on the Prohibition of the Development, Production, Stockpiling and Use of Chemical Weapons and on Their Destruction (CWC): Annexes and Original Signatories, http://www.state.gov/t/avc/trty/175492.htm. Bureau of Arms Control, Verification and Compliance. Retrieved January 19, 2012.

[2] Disarmament lessons from the Chemical Weapons Convention, http://thebulletin.org/web-edition/op-eds/disarmament-lessons-the-chemical-weapons-convention.

[3] Veterans at Risk: Health Effects of Mustard Gas and Lewisite. Committee to Survey the Health Effects of Mustard Gas and Lewisite. Washington, DC: National Academy of Sciences, Institute of Medicine, National Academy Press; 1993.
[4] Sim VM. Chemicals used as weapons in war. Drill's pharmacology in medicine. Mac Graw Hill Book Company; 1971. p. 1232–45.
[5] Boursnell JC, Cohen JA, Dixon M, Francis GE, Greville GD, Needham DM, et al. Studies on mustard gas (betabeta'-dichlorodiethyl sulphide) and some related compounds: 5. The fate of injected mustard gas (containing radioactive sulphur) in the animal body. Biochem J 1946;40(5–6):756–64.
[6] Davison C, Rozman RS, Smith PK. Metabolism of bis-beta-chloroethyl sulfide (sulfur mustard gas). Biochem Pharmacol July 1961;7:65–74.
[7] Porheydari G. Chemical weapon, toxicology and treatment. Tabib Inc.: Baqiyatallah University of Medical Sciences; 2003.
[8] Vijayaraghavan R, Kulkarni A, Pant SC, Kumar P, Rao PV, Gupta N, et al. Differential toxicity of sulfur mustard administered through percutaneous, subcutaneous, and oral routes. Toxicol Appl Pharmacol January 15, 2005;202(2):180–8.
[9] Sharma M, Vijayaraghavan R, Ganesan K. Comparison of toxicity of selected mustard agents by percutaneous and subcutaneous routes. Indian J Exp Biol 2008;46:822–30.
[10] Lakshmana Rao PV, Vijayaraghavan R, Bhaskar AS. Sulphur mustard induced DNA damage in mice after dermal and inhalation exposure. Toxicology 1999;139:39–51.
[11] Marrs TC, et al. In: Chemical warfare agents: toxicology and treatment, 2nd ed. Chichester, West Sussex, England: John Wiley & Sons, Ltd; 2007.
[12] Ghabili K, Agutter PS, Ghanei M, Ansarin K, Panahi Y, Shoja MM. Sulfur mustard toxicity: history, chemistry, pharmacokinetics, and pharmacodynamics. Crit Rev Toxicol May 2011;41(5):384–403.

CHAPTER

3

Biochemical and Cellular–Molecular Mechanisms of Injury From Mustard Gas

Knowledge of the pathophysiology of different diseases is essential for efficient prevention, accurate diagnosis, and effective treatment; therefore, knowledge of the biochemical mechanism of mustard gas or sulfur mustard is of great importance. This chapter will cover the acute phase of injury in addition to its chronicity. The information provided in the chapter will help the readers to use evidence-based medicine in the diagnosis and treatment of sulfur mustard lesions.

Although mustard gas has been recognized as a chemical weapon since 1917, its function has not been fully evaluated yet and, as a result, no definite treatments have yet been found for its complications. In recent decades, scientists have undertaken extensive research on the cellular, molecular, and biochemical effects of mustard gas in the acute phase. As for the mechanism of the injury in the chronic phase, different studies have been performed in our center. The results of these investigations are presented as four main hypotheses in this chapter.

THE MECHANISM OF INJURY IN THE ACUTE PHASE

Different mustard compounds (sulfur and nitrogen) are alkylating agents whose chemical reactivity is based on the ability to undergo internal cyclization of an ethylene group to form a highly reactive episulfonium ion. This process results in the development of a highly electrophilic cyclic ethylene episulfonium derivative that attacks nucleophilic sites in macromolecules. This reactive electrophile is capable of combining with any of the numerous nucleophilic sites present in macromolecules of the cells. As a result, stable compounds are produced that can affect the intrinsic activity of target macromolecules. Since mustard gas affects a variety of cell functions, including alterations in peptides, proteins, RNA, DNA, and cell membrane compounds (Figs. 3.1 and 3.2), scientists are trying to figure out the most important reactions following exposure to mustard.

Since the first effects of mustard gas start with alkylation of DNA and changes in other macromolecules, these changes end in disorders in the dermoepidermal junction and complete blisters are formed following active inflammatory responses in the affected tissue. Moreover, mustard has cholinergic effects and stimulates muscarinic and nicotinic receptors. In addition, through interference with gene expression, changes in the level of cellular proteins, numerous injury mechanisms, and compromising the immune system, sulfur mustard causes long-term injuries to the lungs, skin, and eyes.

Mustard Lung. http://dx.doi.org/10.1016/B978-0-12-803952-6.00003-4

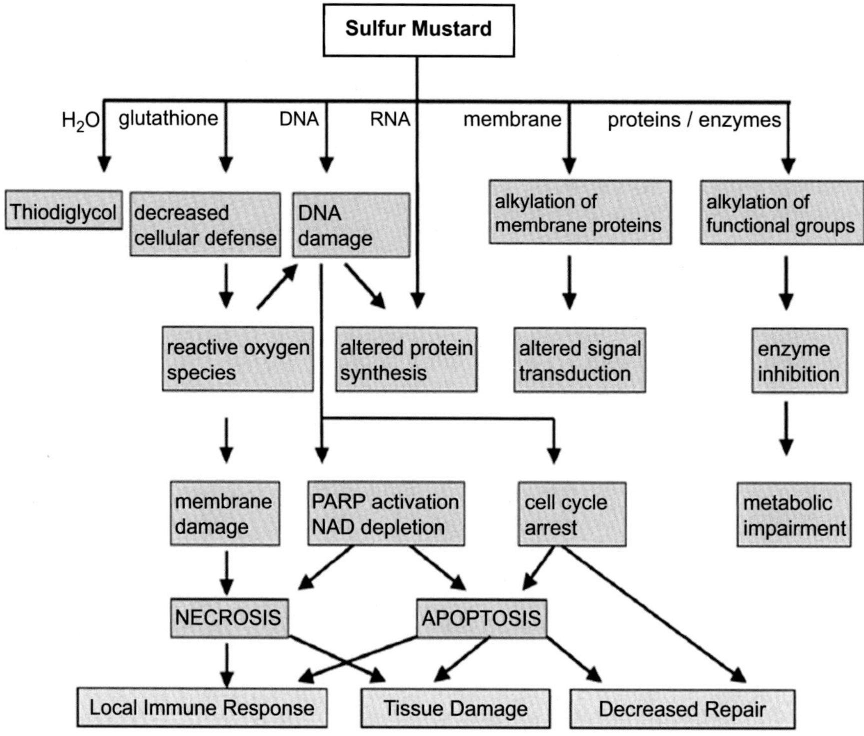

FIGURE 3.1

Reaction of sulfur mustard with different cellular molecules. After the production of episulfonium reactive intermediates, these ions can react with intracellular water, proteins, enzymes, DNA, and RNA. Many intracellular functions are disturbed, resulting in necrosis, apoptosis, inflammation, and disturbances in the cellular regeneration process. *PARP*, poly(adenosine diphosphate-ribose) polymerase.

PATHOPHYSIOLOGY OF MUSTARD GAS INJURY

A variety of injury mechanisms are suggested for sulfur mustard and its pulmonary complications in the chronic phase. Initially, it seemed that the pulmonary involvement in these patients included chronic obstructive pulmonary disease (COPD), asthma, bronchitis, and pulmonary fibrosis. However, lack of appropriate response to specific treatments for these diseases and the pathological and radiological findings indicated special features of the injury in the chemical patients. There are some theories on the development of COPD among which the following are more important and widely recognized [1]:

1. Inflammation
2. Imbalance between oxidative stress and antioxidant mechanisms
3. Apoptosis
4. Proteolysis

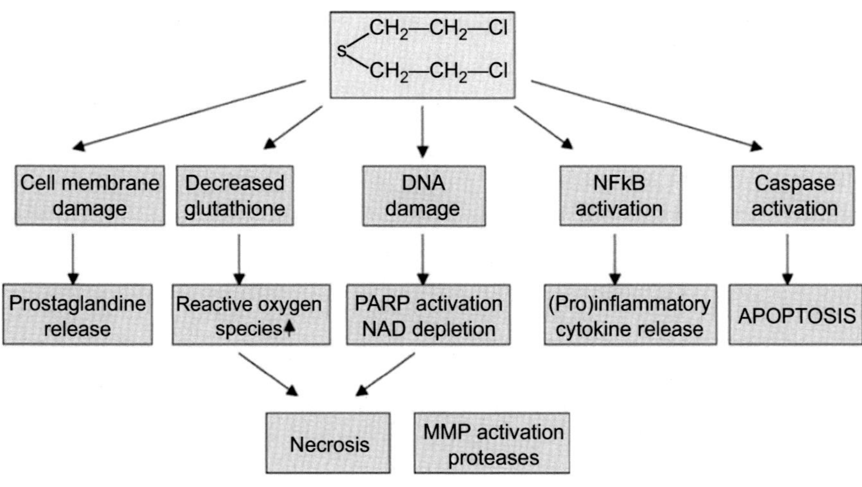

FIGURE 3.2

An example of the theories of the mechanisms of mustard gas injury and the related pathways. Sulfur mustard can damage the DNA molecule directly and indirectly and activates poly(adenosine diphosphate-ribose) polymerase and decreases nicotine adenine dinucleotide, eventually resulting in apoptosis. Moreover, sulfur mustard induces apoptosis through intrinsic and extrinsic pathways. In addition, through releasing inflammatory mediators and cytokines, sulfur mustard results in the release of prostaglandins and activation of NF-kB. It has been shown that sulfur mustard causes overactivation of serine and matrix metalloproteinase.

These mechanisms and their role in the pathogenesis of mustard gas in the chronic phase are further discussed in this chapter.

INFLAMMATION

Different inflammatory mediators, especially interleukin 6 and 8 (IL-6, IL-8), play a major role in the development of asthma and COPD [2]. Moreover, these mediators are recognized as biomarkers to detect these diseases and determine their level of activity. There are increased levels of IL-6 in the serum and bronchoalveolar lavage on the exacerbation of COPD. Inflammatory cells including macrophages, T and B lymphocytes, and neutrophils increase in the pulmonary alveoli and airways of COPD patients. Moreover, inflammatory indexes like C-reactive protein (CRP) are considered prognostic factors of the mortality and severity of the disease [3].

Accordingly, many studies have been performed in chemical patients exposed to mustard gas who have respiratory problems in order to investigate the role of inflammation and related reactions. Since animal studies investigating inflammation in chemical patients have only evaluated the acute phase in vitro, we do not mention their details in this chapter and only address human evidence in real situations that can be used in the diagnosis and treatment of sulfur mustard exposure and have higher external and internal validity.

It is now clear that the main lesion in chemical patients is chronic bronchiolitis obliterans. As the main pulmonary pathology in these patients, the proliferation of fibroblasts and tissue regeneration play the main role in bronchiolitis obliterans. Numerous growth factors can increase the fibroblasts' reaction and collagen accumulation. Transforming growth factor β (TGF-β) is the most important growth factor in this regard. Overproduction of TGF-β in macrophages and endothelial cells can be the cause of the changes resulting from bronchiolitis obliterans. It has been reported that the target cells of TGF-β markedly increase in the bronchoalveolar lavage specimens and target tissues of patients exposed to mustard gas [4].

NF-κB_1 is an important protein group in inflammatory responses. Evaluation of these proteins in chemical patients shows their higher levels in these patients versus the normal population [5].

A study compared the level of High-sensitivity CRP between chemical patients with respiratory complications and the control group. The results showed that the level of this protein was higher in these patients and directly correlated with the severity of the disease [6].

Moreover, the level of cytokines such as IL_8, IL-1β, IL_6, TNF, and IL_{12} was higher in these patients as compared with healthy controls.

Another study reported that the high level of cytokines correlated with the severity of fibrosis [7]. It should be mentioned that the presence of pulmonary fibrosis in chemical patients was suggested in initial studies and was discussed for a long time. More studies have shown that interstitial fibrosis is not an important finding in the pulmonary pathology of these patients. Moreover, it first seemed that inflammation and inflammatory processes, especially together with oxidative stress, played an important role in the pathogenesis of the disease in the early stages of exposure to sulfur mustard, but then it was found that inflammation was not a dominant phenomenon in the chronic phase and developed following oxidative stress. These findings suggested that inflammation did not have a direct and active role during the following years.

The results of the studies performed on chemical victims in Sardasht, Iran, were very interesting. These studies, which evaluated the level of cytokines in outpatients and inpatients, showed that the levels of TNF-α and IL-1β in the sputum of the patients with moderate and severe disease were higher than patients with mild disease while the decrease in fibrinogen was less when compared to advanced cases. In fact, these studies showed that the levels of TNF-α and IL-1β had a positive relationship and the level of fibrinogen had a negative relationship with disease severity [8]. Recently, the high serum level of IL_6 and its relationship with the severity of pulmonary disease has been reported [9].

IL_{19} is a new member of IL_{10} family, which is a proinflammatory cytokine according to some studies and an anti-inflammatory cytokine in some other studies. A study was conducted on 144 victims with severe pulmonary complications, including 57 patients with mild and moderate complications and 55 healthy participants as the control group. This study showed a significant correlation between the serum level of IL_{19} and FEV1/FVC in the exposed group. The mean and median of the serum level of IL_{19} were higher in the control group when compared to the exposure group, and also higher in the mild and moderate exposure group versus the severe exposure group, although the difference was not significant. In the exposure group, the mean and median of serum IL_{19} increased with an increase in FEV1 and FVC and decreased with an increase in FEV1/FVC, but this finding was not also statistically significant [10].

IL_{22} is another member of IL_{10} family that has a proven role in many pulmonary inflammatory diseases. In another study, 219 chemical patients with pulmonary problems as the case group and 54 healthy individuals as the control group were evaluated. Patients in the case group were categorized into severe complications (142 patients) and mild and moderate pulmonary complications (57 patients). The serum level of IL_{22} did not differ significantly between the case and control groups. Moreover, no significant difference was observed in the serum level of IL_{22} between the severe complication and mild to moderate complication groups [11]. More recent studies with a focus on inflammatory markers have shown that not only the level of inflammatory mediators is not high in these patients but also the level of some of them like IL_6, IL_8, CRP, and RF is lower in these patients than the participants in the control group. Moreover, no relationship has been observed between the level of IL_8 and pulmonary symptoms [12]. A study investigated and compared the levels of leptin and adiponectin, as systemic inflammatory markers, but found no difference between chemical patients with pulmonary problems and healthy individuals [13]. These findings indicate that inflammation is not responsible for the chronicity of pulmonary lesions.

It should be noted that although the main pathology in chemical victims is bronchiolitis, this type of bronchiolitis is fundamentally different from obstructive bronchiolitis resulting from pulmonary transplantation. Therefore, the manifestations and response to treatment is different in chemical patients due to the differences between their lungs and the lungs of the patients with other pulmonary problems. Lack of appropriate response to corticosteroid therapy is observed in more than 50% of these patients, indicating that inflammation is not the main pathology. This finding makes active inflammation in these patients even less important.

Evaluation of the open lung biopsy specimens of these patients has shown that even in patients with severe pulmonary complications, there is a mild to moderate lymphocytic infiltration with no trace of considerable interstitial fibrosis, or eosinophilic or neutrophilic inflammation [14].

OXIDATIVE STRESS

Reactive oxygen species (ROS) are very chemically active compounds that are produced in a natural process during aerobic metabolism reactions. Some examples of these compounds are hydrogen superoxide and peroxide that react with different intracellular targets including lipids, proteins, and DNA [15]. Although free radicals are produced during aerobic metabolism, their biological effects on intracellular targets depend on their concentration. Their levels increase during stress. Although increased ROS levels are cytotoxic, lower levels are required for the regulation of certain key physiologic mechanisms such as cell differentiation [16], apoptosis [17], cell proliferation [18], and cell signals transduction [19]. However, their increased levels can result in several types of damage, including cell death, mutations, chromosomal disorders, and carcinogenesis. These free radicals are eliminated from the body through the antioxidant system [20]. The antioxidant system against ROS is divided into enzymatic and nonenzymatic parts. The enzymatic part includes superoxide dismutase, catalase, glutathione peroxidase, glutathione s-transferase, and thioredoxin, and the nonenzymatic part includes proteins like ceruloplasmin, metallothionein, albumin, transferrin, and ferritin to transfer electrons from metals for trapping free radicals. Catalase that has a high affinity for hydrogen peroxide is responsible for neutralizing hydrogen peroxide to water and cellular oxygen in human serum and tissue [21]. Moreover, superoxide dismutase is a metalloenzyme that is the first natural antioxidant in the defensive chain against the toxicity of superoxide radicals. This enzyme catalyzes the conversion of the superoxide radical to hydrogen peroxide and molecular oxygen [22]. In normal individuals with an intact respiratory system, there is a balance between oxidants and antioxidants [23]. Any increase in the concentration of oxidants or any decrease in antioxidants disturbs the balance and results in a state known as oxidative stress.

The current main theory is that oxidative stress plays an important role in the pathogenesis of pulmonary disease through direct injury or involvement of molecular mechanisms effective in lung injury [24].

A large body of evidence supports oxidative stress and mediators in COPD patients. Some of the oxidative stress markers in these patients that are the final products of lipid peroxidation include 4-hydroxy-2-nonenal (4HNE), H_2O_2, and isoprostane. The role of oxidative factors is prominent when their activity and effect overcome those of antioxidant factors. The result of such an imbalance is injury to lipids, proteins, and DNA [25]. Moreover, this process of cellular injury causes apoptosis and injury to the pulmonary matrix including elastin and collagen. Oxidative stress results in the oxidation of arachidonic acid and the formation of new types of prostanoid mediators known as isoprostanes.

Isoprostanes can obstruct the bronchi and produce plasma exudate. Moreover, oxidative stress can induce apoptosis in the endothelial and epithelial cells, indicating the positive feedback between apoptosis and oxidative stress [26]. Oxidative stress exerts its effects through inactivation of antiproteases like X1 antitrypsin or leukoprotease secretion inhibitors or activation of metalloprotease oxidants. Moreover, oxidants have an important role in inflammatory lung injury through the translation of proinflammatory genes [27]. In addition, there is a relationship between response failure to corticosteroid therapy and oxidative stress in COPD patients. Oxidative stress disrupts the expression of glucocorticoid receptors and therefore causes a disorder in the process of transferring these receptors from the cytoplasm to the nucleus [28,29].

The role of oxidative stress in chemical victims exposed to sulfur mustard has been evaluated in different studies. Neutrophils and lymphocytes start to produce oxidant factors following exposure to mustard gas, and these ROS play the main role in the pathogenesis of bronchiolitis obliterans [30,31].

Superoxide dismutases, as a group of potent antioxidants, have a protective role and act as the first lung protective system against free radicals to restore normal cellular metabolism [32]. Contact with sulfur mustard considerably inhibits the activity of superoxide dismutase, glutathione, and catalase [33].

In a study of 250 chemical patients, the specific activity of catalase and superoxide dismutase was measured in the plasma using Paoletti and Cohen methods, respectively, and compared with healthy controls. The results showed that the levels of superoxide dismutase and catalase were higher in patients than controls. Moreover, the serum levels of these enzymes were higher in patients with more severe injury than those with moderate injury. This finding also confirmed the imbalance between oxidants and antioxidants in these patients [34]. The increase in catalase indicates the increase in the concentration of hydrogen peroxide and the presence of oxidative stress [20]. In this study, the increase in the serum level of catalase in chemical patients with late pulmonary complications could result from an increase in hydrogen peroxide. It is important to confirm these hypothesis since increased levels of hydrogen peroxide could lead to the inactivation of the enzyme responsible for this reaction, ie, superoxide dismutase [35].

These cellular events have also been evaluated at the level of gene expression. Superoxide dismutase (SOD) is encoded by different genes, and genetic variations have an important role in the pathogenesis of free radicals [36]. Three distinct types of SODs have been identified in mammals; two isoforms have copper (Cu) and zinc (Zn) in their catalytic center and are located either in the cytoplasm (Cu/Zn SOD) or in the extracellular elements [37,38].

The third isoform, MnSOD (SOD-2), which contains manganese, exists in the mitochondria of aerobic cells [39]. These three isoforms are abundantly found in the human lung under normal conditions. MnSOD is also found in moderate amounts in the respiratory epithelium, type 1 alveolar cells, alveolar macrophages, and interstitial fibroblasts of rats under hypoxic conditions [40]. MnSOD and CuZnSOD are the main superoxide scavengers in the cytosols and mitochondria. In all cases, CuZnSOD exists in the lung as a prominent superoxide dismutase [41].

Considering the role of free radicals in lung injury of sulfur mustard and decreased levels of superoxide dismutases in the respiratory epithelium of pulmonary patients like asthma, Mirbagheri et al. investigated the effect of mustard gas on the level of mRNA and CuZnSOD and MnSOD in the biopsy specimens of the epithelium of the chemical victims. For this reason, 20 patients and 10 healthy controls were evaluated and superoxide dismutases were measured through RT-PCR and immunohistochemistry. The results showed that the level of mRNA related to MnSOD and CuZnSOD increased up to three times in the epithelium of the injured participants as compared with healthy controls while CuZnSOD protein decreased [42]. The disproportion between gene expression and the amount of produced protein indicates a change in the efficacy of gene translation and posttranslational regulations [43].

Glutathione (GSH) and antioxidant enzymes (SOD, catalase (CAT), and glutathione peroxidase) are free radical and oxygen radical scavengers and protect the cells against injury. The leakage of single electrons from the bacterial respiratory chain has been observed with decreased GSH concentrations [44]. This process may start free radical chain reactions resulting in membrane lipid peroxidation. As a result, membrane fluidity decreases and the permeability of the membrane to the ion deactivation of several membrane enzymes increases, leading to the breakage of DNA strands. Malondialdehyde (MDA), which results from breaking of lipid peroxide radicals, is an indicator of oxidative stress. MDA itself can increase the oxidative activity through protein oxidation. Therefore, it could be stated that MDA is both a product and a factor of oxidative stress [45].

These reactions all take part in apoptosis. Under normal conditions, the cells are protected against the toxicity of oxygen radicals through the synergistic action of antioxidation enzymes and GSH. Since the levels of antioxidation enzymes increase in response to the formation of oxygen radicals, determination of their levels in target tissues following exposure to mustard gas may indicate the production of increased amounts of oxygen radicals. Moreover, these compounds can distinguish the potential target tissue in terms of protection against the toxicity of sulfur mustard. The spectrum of the changes of antioxidation enzymes due to sulfur mustard exposure is not yet fully understood.

The levels of glutathione and MDA have also been investigated in chemical patients in order to evaluate oxidative stress. The results have shown that chemical patients with moderate to severe injury had lower levels of glutathione and higher levels of MDA. The increase in MDA indicates increased lipid peroxidation and is a result of free radical production after exposure to sulfur mustard. However, the decrease in glutathione is not limited to pulmonary patients exposed to sulfur mustard as its level also decreases after exposure to other toxins such as ozone and tobacco.

Glutathione S-transferase (GST) plays a role in many cellular functions. Its most important function is to maintain the balance of oxidative stress in the body. Human GST is found in the cytoplasm, mitochondria, and the nucleus. GSTP1 is the most abundant human GST that is mainly found in the lung [46]. In one study, the specimens obtained during the bronchoscopy of the chemical patients were evaluated with mRNA and immunohistochemistry. The results showed that the expression of isoforms of GSTA1, GSTM1, and GSTP1 increased by 2.5, 2.8, and 5.6 times when compared to the normal group, respectively. Immunohistochemistry also showed that the protein expression of GSTP1 was strongly localized in the luminal border of epithelial cells of healthy individuals, while despite the gene expression in the protein level in the patients, it was not found in the epithelial cells of the airways [47].

Recently, the antioxidant activity of paraoxonase-1 (PON1), as an enzyme that prevents the oxidation of lipoproteins during oxidative stress, has received attention. With the same objective, PON1 192 polymorphism and the activity of paraoxonase and arylesterase were evaluated in the sera of the chemical patients. In this study, 101 patients exposed to mustard gas were selected and categorized as mild, moderate, and severe based on their signs and symptoms. The severe decrease in the activity of paraoxonase and arylesterase was proportionate to the severity of the disease.

Considering the PON1 192 polymorphism, it was seen that the RR genotype was more frequent in patients with more severe injury. Moreover, it was noted that only apoA1 protein was measurable in the lavage fluid of the patients with no trace of PON1. It seems that oxidative stress in chemical patients with pulmonary complications is associated with the decreased enzymatic activity of PON1 and increased RR genotype and is positively correlated with the disease severity [48]. Other studies on mustard gas victims have also confirmed the decreased enzymatic activity of PON1 and the relationship with the disease severity of chemical patients [49,50].

Neutrophil gelatinase-associated lipocalin 2 (NGAL) is a group of lipocalins with a protective role against oxidative stress. Evaluation of the expression of NGAL in the airway epithelial cells of chemical patients and healthy controls has shown that mRNA expression in NGAL in patients is about 10 times more than in healthy people. This finding also confirms the active oxidative stress process in these patients that is still observed three decades after exposure to mustard gas [51]. Several studies have suggested an antioxidant effect of apolipoprotein (Apo) A1 and S100 calcium-binding proteins. The high levels of these proteins indicate an imbalance between oxidants and antioxidants in chemical patients. Mehrani et al. used the proteomic method to identify the different proteins expressed in chemical victims in comparison with healthy controls. The results showed that $APOA_1$ was present in the pulmonary lavage fluid of all exposed patients while it was not detected in healthy controls. Moreover, there was a correlation between the severity of pulmonary disease and $APOA_1$ and its isoforms haptoglobulins. S_{100} protein was also detected in all patients with moderate to severe lung injury [52].

A study was performed to determine the phenotypes and plasma activity of alpha 1-antitrypsin in chemical patients with late complications of mustard gas. This historical cohort study was conducted on 100 chemical victims of the Iran–Iraq war and 50 healthy nonsmoker participants with no history of exposure to sulfur mustard (control group). The phenotype of alpha 1-antitrypsin and the trypsin inhibitory capacity was measured. All the participants in this study had the normal MM phenotype while the mean trypsin inhibitory capacity was more in the controls versus the patients. This study showed that alpha 1-antitrypsin activity was decreased in chemical patients due to the effect of oxidative stress [53].

All the evidence is in favor of the effect of the oxidants as the responsible agents for the chronicity of pulmonary lesions in mustard gas victims and emphasizes that future treatments should be directed in this way. Clinical studies using antioxidants have improved clinical and laboratory indexes in these patients. The evidence of the oxidative stress of sulfur mustard in human and animals studies is presented in Table 3.1.

APOPTOSIS

New information suggests the role of apoptosis as one of the most important mechanisms of pathogenesis in COPD. It should be noted that there are two main apoptotic pathways known as the intrinsic and extrinsic apoptotic pathways. The intrinsic pathway is defined by the permeabilization of the mitochondria and cytochrome C release

Table 3.1 Examples of the Evidence of Sulfur Mustard Lung Injury Due to Oxidative Stress Obtained From Cellular, Animal, and Human Studies

References	Effects	System	Vesicant
In Vitro Studies			
[54]	↓ thioredoxin reductase activity	A549 type II human alveolar epithelial cells	CEES
[55]	↑ mitochondrial ROS, ↑ total GSH, ↑ DNA oxidation (8-OHdG), ↓ mitochondrial membrane potential	Human lung epithelial cells and bronchial epithelial cells	CEES
In Vivo Studies			
[56]	↑ CuZn-SOD, n.c. in Mn-SOD, ↓ EC-SOD activity	Guinea pigs (IT)	CEES
[57]	↑ GST activity in lung	Mice (IP)	CEES
[58]	↑ GAPDH, GST activity, ↑ lipid peroxidation, ↑ oxidized GSH	Mice (SQ)	BCS
[59]	↑ SOD, catalase, glutathione peroxidase, glutathione reductase, GST, GAPDH, ↓ reduced GSH, ↑ oxidized GSH, ↑ lipid peroxidation	Mice (SQ)	CCBS
[60]	↓ GSH, ↑ lipid peroxidation, abnormal lung function	Humans (field exposure)	SM
[61]	↑ lipid peroxidation, ↑ iNOS activation	Rats(IT)	NM
[62]	↓ glutathione peroxidase, ↑ iNOS, ↑ lipid peroxidation	Rats(IT)	NM
Antioxidant Treatments			
[63]	↓ GSH – restored by Trolox, Quercetin, GSH ↑ lipid peroxidation – reduced by antioxidants	Mice (inhalation)	SM
[64]	↓ AP-1, c-fos, c-jun, cyclin D1/PCNA, ↓ inflammation and neutrophil infiltration with liposomes containing tocopherols + N-acetylcysteine (NAC)	Guinea pigs (IT)	CEES
[33]	↓ lung injury by NAC pretreatment	Guinea pigs (IT)	CEES
[65]	↓ lung injury with NAC, catalase, resveratrol, DMSO, dimethyl urea pretreatment ↓ lung injury with NAC post-treatment	Rats (intrapulmonary)	CEES
[66]	↓ lung injury with liposomes containing SOD, catalase or NAC, GSH, tocopherol, resveratrol posttreatment instillation	Rats (intrapulmonary)	CEES
[67]	↓ lipid peroxidation and hydroxyproline levels with liposomes containing NAc and tocopherol posttreatment instillation	Guinea pigs (intrapulmonary)	CEES
[61]	↓ iNOS activation and lung damage with iNOS inhibitor and peroxynitrite scavenger	Rats(IT)	NM
[62]	↑ CuZn-SOD, glutathione peroxidase, and iNOS activity, ↓ lipid peroxidation with melatonin pre- and posttreatment	Rats(IT)	NM
Humans Field			
[68]	↓ dyspnea, cough sputum and improved spirometry readings with oral NAC treatment	Humans (field exposure)	SM
[60]	↓ Serum level of GSH and ↑ level of MDA in SM exposed patients compared to nonexposed patients.	Humans (field exposure)	SM

into the cytoplasm. Then, cytochrome C forms a multiprotein complex (apoptosome) that initiates the caspase cascade through caspase-9 [69]. The extrinsic pathway is activated through the receptors of cell death like tumor necrosis factor receptor 1 (TNFR1) and Fas/CD95 ligand available on the cytoplasmic membrane. When the ligands bind to the receptors, death-inducing signaling complex (DISC) is produced that initiates the caspase cascade through caspase-9. Caspase proteins play a pivotal role in the apoptosis process. These proteins include initiator caspases (caspases 2, 8, 9, 10) and effector caspases (caspases 3, 6, 7). Initiator caspases activate effector caspases, which then undergo structure change and carry out the cell death program [70]. The extrinsic pathway is activated by ligand-activated death receptors such as Fas ligand (FasL)—Fas [71]. The binding of Fas-FasL activates cysteine proteases known as caspases that recognize aspartate at their substrate cleavage site and induced apoptosis. Both pathways finally activate caspase-3 that carries out the final stages of apoptosis [71].

It is very important to remember that apoptosis is not an isolated process in the pathogenesis of COPD, and other pathways like oxidative stress further add to its complexity. Apoptosis is a useful process for scavenging neutrophils in inflammatory processes that is present wherever there is oxidative stress in the lung, indicating a positive relationship between them. On the other hand, efferocytosis is a process in which apoptosed cells are scavenged by phagocytes. If this process does not occur, apoptosed neutrophils act as new inducers of oxidative stress. Oxidants inhibit efferocytosis and antioxidants increase it. TGF-β is one of the compounds that increase the efficacy of efferocytosis in the lung. A defect in efferocytosis is observed in many lung diseases like asthma and COPD. In vivo and in vitro studies have confirmed the role of apoptosis as one of the main pathogens in the lung injury of chemical patients and have shown that both intrinsic and extrinsic pathways of apoptosis are active in the lungs of chemical victims.

It has been revealed that there are different types of TGF-β related translations and therefore a high level of TGF-β protein in the lavage fluid of chemical patients (measured by the ELISA method). Moreover, it has been concluded that TGF-β may be responsible for the airway remodeling, homeostasis, and slow progression of the disease in chemical patients.

Three isoforms of TGF-β were evaluated with RT-PCR in one study. The results showed that the levels of TGF-β_1 and TGF-β_3 were considerably higher in chemical patients than healthy controls while TGF-β_2 showed no significant difference [72]. As a result, it was suggested that TGF-β_1 and TGF-β_3 could improve efferocytosis and play a major role in the airway remodeling of these patients. In fact, engulfment of apoptotic cells by macrophages results in the increased release of TGF-β, which induces efferocytosis, inhibition of inflammation and other immunologic reactions, and peroxidation of epithelial and endothelial cells, and finally restoration of the natural pulmonary structure. On the other hand, in other diseases like COPD and cystic fibrosis, the level of TGF-β is lower than normal, resulting in ineffective efferocytosis. These capabilities of TGF-β are a factor for the longer life span of chemical patients when compared with patients with bronchiolitis obliterans resulting from pulmonary transplantation. Recently, a study was conducted to evaluate the expression of TGF and fibroblast receptors of chemical patients' airways. For this reason, the fibroblasts obtained from the airways were cultured and the expression of TGF-β_1, TGF-β_2, TbR-I, and TbR-II genes was evaluated in the fibroblasts of the patients and controls through RT-PCR. Western blotting was also used to measure the levels of the proteins. The results showed that the expression levels of TGF-β_1, TGF-β_2, TbR-I, and TbR-II genes were all increased in the fibroblasts of the patients versus controls. This finding was more prominent for TGF-β_1 [73].

Rosenthal et al. reported that in these patients, exposure of lung cells to low concentrations of sulfur mustard affected the activity of caspase in apoptosis and FLIP protein, which is a caspase-8-like protein [74]. Under normal conditions, this protein prevents oligomerization of caspase-8 and its activation. Moreover, inhibitors of apoptosis proteins (IAPS) like X-linked inhibitor of apoptosis protein (XIAP) inhibit them through binding to caspases 3, 7, and 8. These proteins probably decrease following exposure to sulfur mustard, as well.

Further studies have shown that apoptosis does not occur completely in chemical patients; for example, the level of caspase-3 is not significantly different between chemical patients and the control group. Analysis of the pulmonary lavage fluid with Annexin V-FITC has revealed that major parts of cells have undergone necrosis and a few have completed the apoptosis process.

As mentioned earlier, the low level of glutathione is an important factor to induce apoptosis in chemical patients. Moreover, the homeostasis of calcium and S_{100} protein is decreased in these patients, which plays an important role in the regulation of apoptosis and tissue regeneration.

However, more studies are required to elucidate whether apoptosis is a phenomenon independent of oxidants or is caused due to the effects of oxidants.

INCREASED PROTEOLYTIC ACTIVITY

In COPD patients, any disturbance in the balance between proteolytic and antiproteolytic molecules results in the overactivation of proteolytics. The result of this process is the destruction of the healthy pulmonary parenchyma and the development of emphysema in COPD patients. Since emphysema has not been detected in the lung of nonsmoker chemical patients [14], the presence of proteolytic activity without emphysema is not possible in these patients.

DNA Damage

It has been shown that DNA damage plays an important role in the harmful effects of sulfur mustard. In this regard, genotoxic changes have an important role in the cell death resulting from acute exposure to mustard gas and are therefore responsible for the destruction of target organs [75,76]. Moreover, sulfur mustard is known as a carcinogen in humans due to its devastating effects on DNA, especially in long-term exposure and probably after acute exposure [77,78]. It seems that evaluation of the DNA changes following exposure to sulfur mustard helps to better understand its toxicity [79].

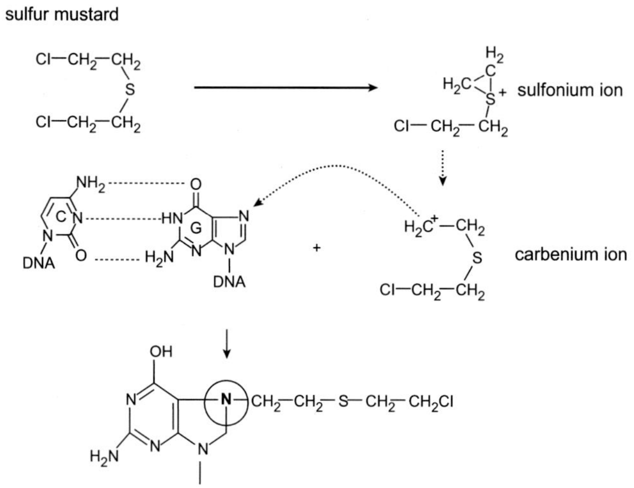

FIGURE 3.3

Reaction of mustard as with DNA structure. After the formation of active mediators including sulfonium and carbonium ions, carbonium reacts with the N7 position of guanine.

Sulfur mustard is a potent alkylating agent that produces most of the DNA damage through the formation of DNA adducts (Fig. 3.3). The major identified DNA adducts are N7-(2-hydroxyethylthioethyl)-guanine, N3-(2-hydroxyethylthioethyl)-adenine, and a cross-linked product, di-(2-guanin-7-yl)ethyl sulfide, which account for only 10–20% of the total adducts.

One of the characteristics of SM-DNA adducts is the instability of N–glycoside bonds. As a result, depurination reaction results in the production of free radicals [80].

In molecular genetics, a DNA adduct is a fragment of DNA that is bonded to a carcinogenic chemical compound in a covalent fashion. This process can initiate the development of a cancerous cell or a carcinogen. DNA adducts are used as biomarkers for confirming exposure to chemicals in research experiments and are also used as a quantitative method to measure the amount of exposure [81].

ROLE OF IMMUNE CELLS IN CHEMICALLY INJURED PATIENTS

In the acute and chronic phase after exposure to sulfur mustard, different changes occur in the blood cells. The results of studies on the homeostasis of immune cells in chemical patients are sometimes contradictory. An overview of the studies follows.

In the acute phase of the disease and after the first exposure to sulfur mustard, transient leukocytosis and leukopenia are observed that return to normal within 4 weeks [82]. Moreover, lethal doses of sulfur mustard cause considerable cytopenia and increase the patient's susceptibility to secondary infections [83]. In the acute phase of exposure to sulfur mustard, a tangible decrease is noted in most hematopoietic cells that return to normal in the chronic phase.

GRANULOCYTES

There are reports of increased neutrophils and eosinophils in the bronchoalveolar lavage fluid of the patients 10 years after exposure to sulfur mustard, and a strong correlation has been observed between the number of these cells and the amount of fibrosis [84]. However, no difference was found in peripheral blood neutrophils and eosinophils between pulmonary patients with mustard gas exposure and normal individuals [83]. In fact, neutrophilic alveolitis is the dominant phenomenon and neutrophils and eosinophils are dominant cells in the bronchoalveolar lavage fluid of the patients.

MONOCYTES AND MACROPHAGES

There are different reports of the population of monocytes and macrophages in the peripheral blood and bronchoalveolar lavage fluid of the chemical patients. Pourkaveh et al. reported that the number of white cells increased with the increase in the severity of the disease, and the increase was more noticeable in the severe disease group versus the control group. Moreover, the percentage of $CD14^+$ and $CD14^+/CD16^+$ cells showed no significant difference between patients and controls, while the percentage of $CD14^+/HLA\text{-}DR^+$ cells showed a significant increase in the moderate disease group versus the control group and decreased considerably in the severe disease group. In this study, it was finally concluded that after years, the trend of monocyte production in the bone marrow was intact and only their function was disturbed. The authors believed that the increase in white blood cells in the severe group resulted from chronic infections and lung injury in these patients [85]. My colleagues and I studied peripheral blood of the chemical patients exposed to mustard gas but found no difference between them and normal individuals, while Mahmoudi et al. reported a significant increase in monocytes and macrophages in chemical patients' peripheral blood 16–20 years after exposure to mustard gas as compared with normal people.

Decreased NK cells in sulfur mustard–exposed pulmonary patients have been suggested as a main reason for viral infections in these patients. One study on the peripheral blood of these patients showed a significant decrease in NK cells when compared to healthy subjects [86]. Moreover, a decrease in NK CD25+ cells was observed in the peripheral blood of the pulmonary patients exposed to sulfur mustard, but their activity was increased in patients with more advanced disease [87]. A more recent study reported an increase in NK cells of the peripheral blood of these patients [88].

T CELLS

An increase in Tc1 cells has been reported in the workers of chemical warfare companies. Moreover, a decrease in the ratio of TCD4+ to TCD8+ cells and an increase in Th cells have been reported in the bronchoalveolar lavage fluid of the patients with bronchiectasis resulting from exposure to sulfur mustard. The decrease in Th cells is more severe in patients with more advanced disease [88]. In comparison with healthy individuals, TCD3+ cells increase significantly in the peripheral blood of the chemical patients [86]. A study by Shaker et al. revealed the normal percentage of CD45+ cells while the percentage of T-helper and cytotoxic T cells was decreased in chemical patients with severe disease versus mild disease. Moreover, the results showed that CD4+/CD25+ cells were heavily affected in most patients and had an increase in comparison with other patients [89]. Furthermore, there are reports of increased $CD3^{+}CD16^{+}CD56^{+}$ NKT-like cells in the peripheral blood of the patients [88].

In another study, F344 rats and cynomolgus monkeys were exposed to sulfur mustard inhalation. In rats, within 72 h, inflammatory cytokines and chemokines including IL-1β, TNF-α, IL-2, IL-6, CCL2, CCL3, CCL11, and CXCL1 flooded the lungs along with neutrophilic infiltration. After about two weeks on (the chronic phase), lymphocytic infiltration and increased expression of cytokines and chemokines continued; moreover, TGF-β, which was not detectable in the acute phase, substantially increased in the chronic phase. This condition continued until the animals' death. In addition, the chronic phase was associated with myofibroblast proliferation, collagen deposition, and the presence of IL-17(+) cells. After 30 days, inhalation of sulfur mustard induced the aggregation of IL-17(+) cells in the inflamed sections of the monkeys' lungs [90].

CHANGES IN ANTIBODIES

A study showed that only IgM and IgG4 decreased in the sera of sulfur mustard patients. Another study on 372 male patients 20 years after contamination with sulfur mustard showed no change in the concentration of other antibodies [73]. An investigation found an increase in the concentration of IgG with no change in the concentration of other isotypes in the bronchoalveolar lavage fluid of the mustard gas–induced bronchiectasis patients [91]. Keyhani et al. reported that the concentration of IgG and IgA increased until one month after exposure to sulfur mustard while the concentration of IgM did not change [92]. Moreover, Ahmadi et al. concluded that the concentration of the antibodies had no effect on the amount of antibodies and decreased gradually after one month [93].

Because of the limited number of the studies in this regard, it is difficult to make a definite conclusion on the exact role of B cells in the pathogenesis of sulfur mustard–induced lung injuries; however, it can be generally stated that antibodies do not play an important role in the pathophysiology of the chronic phase of the disease.

REFERENCES

[1] Ghanei M, Harandi AA. Molecular and cellular mechanism of lung injuries due to exposure to sulfur mustard: a review. Inhal Toxicol June 2011;23(7):363–71.

[2] Danilko KV, Korytina GF, Akhmidishina LZ, Ianbaeva DG, Zagidullin S, Victorova TV. Association of cytokines genes (ILL, IL1RN, TNF, LTA, IL6, IL8, IL0) polymorphicmarkers with chronic obstructive pulmonary disease. Mol Biol (Mosk) 2007;41(1):26–36.

[3] Higashimoto Y, Iwata T, Okada M, Satoh H, Fukuda K, Tohda Y. Serum biomarkers as predictors of lung function decline in chronic obstructive pulmonary disease. Respir Med 2009;103(8):1231–8.

[4] Aghanouri R, Ghanei M, Aslani J, Keivani-Amine H, Rastegar F, Karkhane A. Fibrogenic cytokine levels in bronchoalveolar lavage aspirates 15 years after exposure to sulfur mustard. Am J Physiol Lung Cell Mol Physiol 2004;287:L1160–4.

[5] Yazdani S, Karimfar MH, Imani Fooladi AA, Mirbagheri L, Ebrahimi M, Ghanei M, et al. Nuclear factor κB1/RelA mediates the inflammation and/or survival of human airway exposed to sulfur mustard. J Recept Signal Transduct Res October 2011;31(5):367–73.

[6] Attaran D, Lari SM, Khajehdaluee M, Ayatollahi H, Towhidi M, Asnaashari A, et al. Highly sensitive C-reactive protein levels in Iranian patients with pulmonary complication of sulfur mustard poisoning and its correlation with severity of airway diseases. Hum Exp Toxicol 2009;28(12):739–45.

[7] Emad A, Emad Y. CD4/CD8 ratio and cytokine levels of the BAL fluid in patients with bronchiectasis caused by sulfur mustard gas inhalation. J Inflamm (Lond) 2007;4:2.

[8] Yaraee R, Hassan ZM, Pourfarzam S, Rezaei A, Faghihzadeh S, Ebtekar M, et al. Fibrinogen and inflammatory cytokines in spontaneous sputum of sulfur-mustard-exposed civilians–Sardasht-Iran Cohort Study. Int Immunopharmacol November 2013;17(3):968–73.

[9] Shohrati M, Harandi AA, Najafian B, Saburi A, Ghanei M. The role of serum level of interleukin-6 in severity of pulmonary complications of sulfur mustard injuries. Iran J Med Sci July 2014;39(4).

[10] Ghazanfari T, Sajadi M, Kavandi E, Yarai R, Poorfarzam S, Rezai A, et al. Assessment of Serum level of IL-19 in chemical victims with pulmonary complications. IJWPH 2011;3(4):6–14.

[11] Ghazanfari T, Kavandi E, Sajjadi M, Poorfarzam S, Rezaii A, Sharifnia Z, et al. Serum level of IL-22 in chemical victims with pulmonary complications. IJWPH 2013;5(4):1–8.

[12] Pourfarzam S, Ghazanfari T, Yaraee R, Ghasemi H, Hassan ZM, Faghihzadeh S, et al. Serum levels of IL-8 and IL-6 in the long term pulmonary complications induced by sulfur mustard: Sardasht-Iran Cohort Study. Int Immunopharmacol 2009;9(13–14):1482–8.

[13] Rokni Yazdi H, Lari S, Attaran D, Ayatollahi H, Mohsenizadeh A. The serum levels of adiponectin and leptin in mustard lung patients. Hum Exp Toxicol September 24, 2013;33.

[14] Ghanei M, Tazelaar HD, Chilosi M, Harandi AA, Peyman M, Akbari HM, et al. An international collaborative pathologic study of surgical lung biopsies from mustard gas-exposed patients. Respir Med 2008;102(6):825–30.

[15] Cerutti PA. Prooxidant states and cancer. Science 1985;227:375–81.

[16] Allen RG, Balon AK. Oxidative influence in development and differentiation: and overview of a free radical theory of development. Free Radic Biol Med 1989;6:631–61.

[17] Hockenbery DM, Oltvai ZN, Yin XM, Milliman CL, Korsmeyer SJ. Bcl-2 functions in an antioxidant pathway to prevent apoptosis. Cell 1993;75:241–51.

[18] Shibanuma M, Kuroki T, Nose K. Induction of DNA replication and expression of proto-oncogenes c-myc and c-fos in quiescent Balb/3T3 cells by xanthine/xanthine oxidase. Oncogene 1988;3:17–21.

[19] Lo YYC, Wong JMS, Cruz TF. Reactive oxygen species mediate cytokine activation of c-Jun NH2-terminal kinases. J Biol Chem 1996;271:15703–7.

[20] Rahman I, MacNee W. Role of oxidants/antioxidants in smoking-induced lung diseases. Free Radic Biol Med 1996;21:669–81.

[21] Weydert CJ, Cullen JJ. Measurement of superoxide dismutase, catalase and glutathione peroxidase in cultured cells and tissue. Nat Protoc January 2010;5(1):51–66.

[22] Fridovich I, Freeman B. Antioxidant defenses in the lung. Annu Rev Physiol 1986;48: 693–702.

[23] Sies H. Antioxidant functions of vitamin E, and C, b-carotene, and other carotenoids. Ann N Y Acad Sci 1992;669:7–20.

[24] Adamson IY, Bowden DH. The pathogenesis of bleomycin-induced pulmonary fibrosis in mice. Am J Pathol 1974;77:185–97.

[25] Kirkham PA, Barnes PJ. Oxidative stress in COPD. Chest July 2013;144(1):266–73.

[26] Barnes PJ. Cellular and molecular mechanisms of chronic obstructive pulmonary disease. Clin Chest Med March 2014;35(1):71–86.

[27] Demedts IK, Demoor T, Bracke KR, Joos GF, Brusselle GG. Role of apoptosis in the pathogenesis of COPD and pulmonary emphysema. Respir Res 2006;7:53.

[28] Hutchison KA, Matić G, Meshinchi S, Bresnick EH, Pratt WB. Redox manipulation of DNA binding activity and BuGR epitope reactivity of the glucocorticoid receptor. J Biol Chem June 5, 1991;266(16):10505–9.

[29] Okamoto K, Tanaka H, Ogawa H, Makino Y, Eguchi H, Hayashi S, et al. Redox-dependent regulation of nuclear import of the glucocorticoid receptor. J Biol Chem April 9, 1999;274(15):10363–71.

[30] Ichinose M, Sugiura H, Yamagata S, Koarai A, Shirato K. Increase in reactive nitrogen species production in chronic obstructive pulmonary disease airways. Am J Respir Crit Care Med 2000;162(2 Pt 1):701–6.

[31] Saetta M, Turato G, Maestrelli P, Mapp CE, Fabbri LM. Cellular and structural bases of chronic obstructive pulmonary disease. Am J Respir Crit Care Med 2001;163(6): 1304–9.

[32] Oury TD, Day BJ, Crapo JD. Extracellular superoxide dismutase: a regulator of nitric oxide bioavailability. Lab Invest 1996;75(5):617–36.

[33] Das SK, Mukherjee S, Smith MG, Chatterjee D. Prophylactic protection by N-acetylcysteine against the pulmonary injury induced by 2-chloroethyl ethyl sulfide, a mustard analogue. J Biochem Mol Toxicol 2003;17(3):177–84.

[34] Shohrati M, Ghanei M, Shamspour N, Jafari M. Activity and function in lung injuries due to sulphur mustard. Biomarkers November 2008;13(7):728–33.

[35] Varshavskii BIA, Trubnikov GV, Galaktipmpva LP, Koreniak NA, Koledeznaia IL, Oberemok AN. Oxidant–antioxidant status of patients with bronchial asthma during in halationand systemic glucocortioidtherapy. Ter Arh 2003;75:21–4.

[36] Forsberg L, de Faire U, Morgenstern R. Oxidative stress, human genetic variation, and disease. Arch Biochem Biophys 2001;389(1):84–93.

[37] Marklund SL. Human copper-containing superoxide dismutase of high molecular weight. Proc Natl Acad Sci U S A 1982;79(24):7634–8.

[38] Crapo JD, Oury T, Rabouille C, Slot JW, Chang LY. Copper, zinc superoxide dismutase is primarily a cytosolic protein in human cells. Proc Natl Acad Sci U S A 1992;89(21):10405–9.

[39] Weisiger RA, Fridovich I. Mitochondrial superoxide simutase. Site of synthesis and intramitochondrial localization. J Biol Chem 1973;248(13):4793–6.

[40] Chang LY, Kang BH, Slot JW, Vincent R, Crapo JD. Immunocytochemical localization of the sites of superoxide dismutase induction by hyperoxia in rat lungs. Lab Invest 1995;73(1):29–39.

[41] Comhair SA, Bhathena PR, Dweik RA, Kavuru M, Erzurum SC. Rapid loss of superoxide dismutase activity during antigen-induced asthmatic response. Lancet 2000;355(9204):624.

[42] Mirbagheri L, Habibi Roudkenar M, Imani Fooladi AA, Ghanei M, Nourani MR. Downregulation of super oxide dismutase level in protein might be due to sulfur mustard induced toxicity in lung. Iran J Allergy Asthma Immunol May 15, 2013;12(2):153–60. PubMed PMID: 23754354.

[43] Haq F, Mahoney M, Koropatnick J. Signaling events for metallothionein induction. Mutat Res 2003;533(1–2):211–26.

[44] Dekhuijzen PN. Acetylcysteine in the treatment of severe COPD. Ned Tijdschr Geneeskd 2006;150:1222–6.

[45] Fidan F, Unlu M, Koken T, Tetik L, Akgun S, Demirel R, et al. Oxidant-Antioxidant Status and pulmonary function in welding workers. J Occup Health 2005;47:286–92.

[46] Haddad J, Safieh-Garabedian B, Saade NE, Lauterbach R. Inhibition of glutathione-related enzymes augments LPS mediated cytokine biosynthesis: Involvement of an IKB/NFkB sensitive pathway in the alveolar epithelium. Int Immunopharmacol 2002;2:1567–83.

[47] Nourani MR, Azimzadeh S, Ghanei M, Imani Fooladi AA. Expression of glutathione S-transferase variants in human airway wall after long-term response to sulfur mustard. J Recept Signal Transduct Res December 17, 2013;34.

[48] Golmanesh L, Bahrami F, Pourali F, Vahedi E, Wahhabaghai H, Mehrani H, et al. Assessing the relationship of paraoxonase-1 Q192R polymorphisms and the severity of lung disease in SM-exposed patients. Immunopharmacol Immunotoxicol June 2013;35(3):419–25.

[49] Taravati A, Ardestani SK, Ziaee AA, Ghorbani A, Soroush MR, Faghihzadeh S, et al. Effects of paraoxonase 1 activity and gene polymorphisms on long-term pulmonary complications of sulfur mustard-exposed veterans. Int Immunopharmacol November 2013;17(3):974–9.

[50] Taravati A, Ardestani SK, Soroush MR, Faghihzadeh S, Ghazanfari T, Jalilvand F, et al. Serum albumin and paraoxonase activity in Iranian veterans 20 years after sulfur mustard exposure. Immunopharmacol Immunotoxicol August 2012;34(4):706–13.

[51] Ebrahimi M, Roudkenar MH, Imani Fooladi AA, Halabian R, Ghanei M, Kondo H, et al. Discrepancy between mRNA and protein expression of neutrophil gelatinase-associated lipocalin in bronchial epithelium induced by sulfur mustard. J Biomed Biotechnol 2010;2010:823131.

[52] Mehrani H, Ghanei M, Aslani J, Golmanesh L. Bronchoalveolar lavage fluid proteomic patterns of sulfur mustard-exposed patients. Proteomics Clin Appl 2009;3:1191–200.

[53] Shohrati M, Shamspour N, Babaei F, Harandi AA, Mohsenifar A, Aslani J, et al. Evaluation of activity and phenotype of alpha1-antitrypsin in a civil population with respiratory complications following exposure to sulfur mustard 20 years ago. Biomarkers February 2010;15(1):47–51.

[54] Jan YH, Heck DE, Gray JP, Zheng H, Casillas RP, Laskin DL, et al. Selective targeting of selenocysteine in thioredoxin reductase by the half mustard 2-chloroethyl ethyl sulfide in lung epithelial cells. Chem Res Toxicol June 21, 2010;23(6):1045–53.

[55] Gould NS, White CW, Day BJ. A role for mitochondrial oxidative stress in sulfur mustard analog 2-chloroethyl ethyl sulfide-induced lung cell injury and antioxidant protection. J Pharmacol Exp Ther 2009;328:732–9.

[56] Mukhopadhyay S, Rajaratnam V, Mukherjee S, Smith M, Das SK. Modulation of the expression of superoxide dismutase gene in lung injury by 2-chloroethyl ethyl sulfide, a mustard analog. J Biochem Mol Toxicol 2006;20(3):142–9.

[57] Kim YB, Lee YS, Choi DS, Cha SH, Sok DE. Change in glutathione S-transferase and glyceraldehyde-3-phosphate dehydrogenase activities in the organs of mice treated with 2-chloroethyl ethyl sulfide or its oxidation products. Food Chem Toxicol March 1996;34(3):259–65.

[58] Elsayed NM, Omaye ST, Klain GJ, Korte Jr DW. Free radical-mediated lung response to the monofunctional sulfur mustard butyl 2-chloroethyl sulfide after subcutaneous injection. Toxicology 1992;72:153–65.
[59] Elsayed NM, Omaye ST. Biochemical changes in mouse lung after subcutaneous injection of the sulfur mustard 2-chloroethyl 4-chlorobutyl sulfide. Toxicology 1992;72(2):153–65.
[60] Shohrati M, Ghanei M, Shamspour N, Babaei F, Abadi MN, Jafari M, et al. Glutathione and malondialdehyde levels in late pulmonary complications of sulfur mustard intoxication. Lung January–February 2010;188(1):77–83.
[61] Yaren H, Mollaoglu H, Kurt B, Korkmaz A, Oter S, Topal T, et al. Lung toxicity of nitrogen mustard may be mediated by nitric oxide and peroxynitrite in rats. Res Vet Sci August 2007;83(1):116–22.
[62] Ucar M, Korkmaz A, Reiter RJ, Yaren H, Oter S, Kurt B, et al. Melatonin alleviates lung damage induced by the chemical warfare agent nitrogenmustard. Toxicol Lett September 10, 2007;173(2):124–31.
[63] Kumar O, Sugendran K, Vijayaraghavan R. Protective effect of various antioxidants on the toxicity of sulphur mustard administered tomice by inhalation or percutaneous routes. Chem Biol Interact March 14, 2001;134(1):1–12.
[64] Mukhopadhyay S, Mukherjee S, Stone WL, Smith M, Das SK. Role of MAPK/AP-1 signaling pathway in the protection of CEES-induced lung injury by antioxidant liposome. Toxicology July 10, 2009;261(3):143–51.
[65] McClintock SD, Till GO, Smith MG, Ward PA. Protection from half-mustardgas-induced acute lung injury in the rat. J Appl Toxicol July–August 2002;22(4):257–62.
[66] McClintock SD, Hoesel LM, Das SK, Till GO, Neff T, Kunkel RG, et al. Attenuation of half sulfurmustard gas-induced acute lung injury in rats. J Appl Toxicol March–April 2006;26(2):126–31.
[67] Mukherjee S, Stone WL, Yang H, Smith MG, Das SK. Protection of half sulfur mustard gas-induced lung injury in guinea pigs by antioxidant liposomes. J Biochem Mol Toxicol March–April 2009;23(2):143–53.
[68] Ghanei M, Shohrati M, Jafari M, Ghaderi S, Alaeddini F, Aslani J. N-acetylcysteine improves the clinical conditions of mustard gas-exposed patients with normal pulmonary function test. Basic Clin Pharmacol Toxicol November 2008;103(5):428–32.
[69] Kroemer G, Reed JC. Mitochondrial control of cell death. Nat Med 2000;6:513–9.
[70] Kuwano K. Epithelial cell apoptosis and lung remodeling. Cell Mol Immunol December 2007;4(6):419–29.
[71] Thorburn A. Death receptor-induced cell killing. Cell Signalling 2004;16(2):139–44.
[72] Zarin AA, Behmanesh M, Tavallaei M, Shohrati M, Ghanei M. Overexpression of transforming growth factor (TGF)-beta1 and TGF-beta3 genes in lung of toxic-inhaled patients. Exp Lung Res 2010;36:284–91.
[73] Mirzamani MS, Nourani MR, Imani Fooladi AA, Zare S, Ebrahimi M, Yazdani S, et al. Increased expression of transforming growth factor-β and receptors in primary human airway fibroblasts from chemical inhalation patients. Iran J Allergy Asthma Immunol May 15, 2013;12(2):144–52.
[74] Rosenthal DS, Velena A, Chou FP, Schlegel R, Ray R, Benton B, et al. Expression of dominant-negative Fas-associated death domain blocks human keratinocyte apoptosis and vesication induced by sulfur mustard. J Biol Chem 2003;278:8531–40.
[75] Debiak M, Kehe K, Bürkle A. Role of poly(ADP-ribose) polymerase in sulfur mustard toxicity. Toxicology September 1, 2009;263(1):20–5.

[76] Kehe K, Balszuweit F, Steinritz D, Thiermann H. Molecular toxicology of sulfur mustard-induced cutaneous inflammation and blistering. Toxicology September 1, 2009;263(1):12–9.
[77] Emadi SN, Mortazavi M, Mortazavi H. Late cutaneous manifestations 14 to 20 years after wartime exposure to sulfur mustard gas: a long-term investigation. Arch Dermatol August 2008;144(8):1059–61.
[78] Easton DF, Peto J, Doll R. Cancers of the respiratory tract in mustard gas workers. Br J Ind Med October 1988;45(10):652–9.
[79] Batal M, Boudry I, Mouret S, Cléry-Barraud C, Wartelle J, Bérard I, et al. DNA damage in internal organs after cutaneous exposure to sulphur mustard. Toxicol Appl Pharmacol July 1, 2014;278(1):39–44.
[80] Boysen G, Pachkowski BF, Nakamura J, Swenberg JA. The formation and biological significance of N7-guanine adducts. Mutat Res August 2009;678(2):76–94.
[81] La DK, Swenberg JA. DNA adducts: biological markers of exposure and potential applications to risk assessment. Mutat Res September 1996;365(1–3):129–46.
[82] Anderson DR, Holmes WW, Lee RB, Dalal SJ, Hurst CG, Maliner BI, et al. Sulfur mustard-induced neutropenia: treatment with granulocyte colony-stimulating factor. Mil Med May 2006;171(5):448–53.
[83] Ghanei M. Delayed haematological complications of mustard gas. J Appl Toxicol November–December 2004;24(6):493–5.
[84] Emad A, Emad Y. Increased in CD8 T lymphocytes in the BAL fluid of patients with sulfur mustard gas-induced pulmonary fibrosis. Respir Med April 2007;101(4):786–92.
[85] Pourkaveh S, Mohammad Hasan Z, Mosafa N, Pourhamid S. Evaluation of the immunity of monocytes cell line using peripheral markers in the blood of chemical victims of Iraq-Iran war. J Kermanshah Med University 2003;6(4):1–9.
[86] Mahmoudi M, Hefazi M, Rastin M, Balali-Mood M. Long-term hematological and immunological complications of sulfur mustard poisoning in Iranian veterans. Int Immunopharmacol August 2005;5(9):1479–85.
[87] Ghotbi L, Hassan Z. The immunostatus of natural killer cells in people exposed to sulfur mustard. Int Immunopharmacol June 2002;2(7):981–5.
[88] Ghazanfari T, Kariminia A, Yaraee R, Faghihzadeh S, Ardestani SK, Ebtekar M, et al. Long term impact of sulfur mustard exposure on peripheral blood mononuclear subpopulations–Sardasht-Iran Cohort Study (SICS). Int Immunopharmacol November 2013;17(3):931–5.
[89] Shaker Z, Hassan ZM, Sohrabpoor H, Mosaffa N. The Immunostatus of T Helper and T Cytotoxic cells in the patients ten years after exposure to sulfur mustard. Immunopharmacol Immunotoxicol 2003;25(3):423–30.
[90] Mishra NC, Rir-sima-ah J, Grotendorst GR, Langley RJ, Singh SP, Gundavarapu S, et al. Inhalation of sulfur mustard causes long-term T cell-dependent inflammation: possible role of Th17 cells in chronic lung pathology. Int Immunopharmacol May 2012;13(1):101–8.
[91] Emad A, Rezaian GR. Immunoglobulins and cellular constituents of the BAL fluid of patients with sulfur mustard gas-induced pulmonary fibrosis. Chest May 1999;115(5):1346–51.
[92] Keyhani A, Eslami MB, Razavimanesh H. The short-term effect of mustard gas on the serum immunoglobulin levels. Iran J Allergy Asthma Immunol March 2007;6(1):15–9.
[93] Ahmadi K, Solgue G. Cytokine pattern in sera and broncho-alveolar lavage six months after single exposure to sulfur mustard. Med J Islamic Repub Iran 2006;20(2):52–6.

CHAPTER

4 The Effects of Mustard Gas on Pulmonary Function and Structure

LONG-TERM CLINICAL AND HISTOLOGICAL EVALUATION OF INDIVIDUALS EXPOSED TO MUSTARD GAS

Among the three major organs that are involved after exposure to mustard gas, ie, the lung, eyes, and skin, the lungs are affected the most. Most preliminary studies have evaluated the acute effects of mustard gas; however, more recent studies on the chronic effects and complications of mustard gas have sometimes resulted in contradictory results. Nonetheless, it seems that more investigation on the late complications of mustard gas may correct diagnostic errors and improve diagnostic and treatment interventions. The results of these studies can help to differentiate these patients from other chronic pulmonary patients such as those with asthma, chronic bronchitis, bronchiectasis, interstitial lung disease, and pulmonary fibrosis.

Inhalation of mustard gas can cause different degrees of lung injury. Previous studies reported pulmonary fibrosis following exposure to sulfur mustard. However, our study on the patients with sulfur mustard–induced permanent injuries and their comprehensive evaluation in terms of clinical, radiological, and histological investigations revealed that in contrast to animal models, pulmonary fibrosis is not the dominant pathology in humans. Based on the complementary reports, pulmonary fibrosis is the least significant pulmonary involvement in these patients. It is not clear whether this disease has a picture of interstitial pneumonia like other interstitial lung diseases or its clinical picture is unclear. The findings in this regard are very general and include parenchymal and airways mucosal injury.

More recent studies using high-resolution computed tomography (HRCT) have reported air trapping as a common finding, suggesting bronchiolitis obliterans as the underlying disease. It should be noted that HRCT should be necessarily performed on expiration for the evaluation of bronchiolitis obliterans and air trapping because obstruction in bronchioles is responsible for air trapping, which is best viewed on expiration (Fig. 4.1).

Pulmonary air trapping more than 25% on HRCT strongly suggests bronchiolitis obliterans [1]. The mosaic pattern in bronchiolitis obliterans results from hyperventilation of terminal alveoli to bronchioles due to obstruction. On the other hand, due to vasoconstriction, low perfusion areas and noninvolved areas with normal or even increased perfusion produce a mosaic pattern on HRCT. Although most patients were not in a hyperventilation state, the mosaic pattern was observed on HRCT. Spirometry of these patients shows obstructive, restrictive, or sometimes normal patterns.

Mustard Lung. http://dx.doi.org/10.1016/B978-0-12-803952-6.00004-6

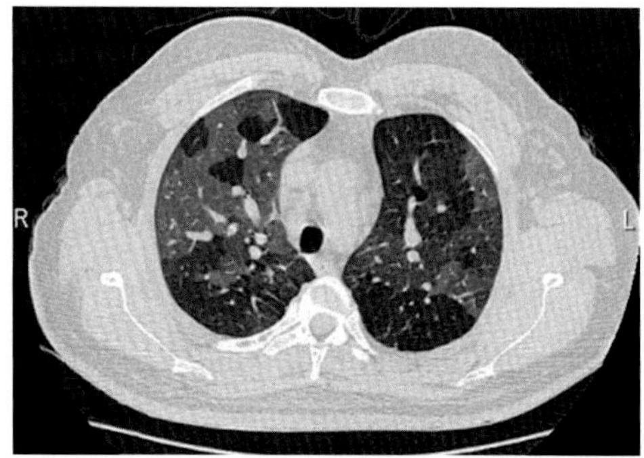

FIGURE 4.1

High-resolution computed tomography (HRCT) of a 43-year-old chemical patient 20 years after exposure to sulfur mustard. Note the expiratory air trapping.

It has been shown that specimens provided by needle biopsy or the transbronchial method do not yield adequate tissue to differentiate the type of bronchiolitis, and the reported pathology is usually organizing pneumonia. If possible, open biopsy of the pulmonary tissue or obtaining a tissue sample through video-assisted thoracoscopic surgery (VATS) provides the diagnostic possibility of bronchiolitis obliterans while bronchoscopic biopsy is an accepted technique for the detection and follow-up of patients with organized pneumonia; for example, this method is used to evaluate chronic transplant rejection in patients receiving pulmonary transplantation. In our study, histologic evaluation of the patients exposed to chemical toxins through bronchoscopic biopsy revealed damage to the tracheobronchial tree, thickening of the basal membrane, edema and infiltration of mononuclear cells to lamina propria, lamina propria fibrosis, and hyperplasia of muscularis mucosa. These characteristics could indicate an inflammatory and chronic reaction following the inhalation of chemicals [2].

CHANGES OF THE RESPIRATORY TRACTS DUE TO MUSTARD GAS

The extent of the changes in the respiratory tracts following contact with mustard gas is determined by the duration of contact and concentration of the chemical agent in the inhaled air. In warm environments, the effects of mustard gas on the respiratory system are augmented. The findings from World War I (WWI) can be interpreted to show the high prevalence of secondary infection in postmortem specimens [3].

UPPER RESPIRATORY TRACT

In cases of severe contact and in the acute phase, the epithelial layer of the larynx, trachea, and bronchi undergo necrosis, and a diphtheritic membrane may form. In cases with less severe contact, hyperemia and petechia are common findings in the superficial layers of the respiratory tract. Similar injuries have also been noted in the respiratory tracts of animals exposed to mustard gas until several months after contact. In WWI, cases of gangrenous changes in the trachea were also reported. Light microscopy shows secretions of the epithelial cells, fibrin, and mucus. In these reports, the basal membrane is not clearly visible due to swelling, and edema in the epithelial tissues along with infiltration of inflammatory cells and vasodilation are common findings.

In cases with more severe contact, injuries extend to the connective tissue and airways smooth muscle. During the recovery phase, extensive metaplasia of the squamous cells is observed and its first changes can be detected in mucosal duct. Metaplastic stratified squamous epithelium completely covers the injured parts [3]. In these studies, the growth stages of these epithelial cells are not fully described and reconstruction of pseudostratified columnar respiratory epithelial cells is not well documented.

TRACHEOBRONCHOMALACIA AND AIR TRAPPING FOLLOWING EXPOSURE TO MUSTARD GAS

Tracheobronchomalacia is one of the major complications of exposure to sulfur mustard. Tracheobronchomalacia and air trapping are always evident on the chest HRCT of mustard gas patients.

The findings of our studies for the first time revealed that air trapping and tracheobronchomalacia were associated as long-term complications of exposure to sulfur mustard. The association of air trapping and bronchiolitis obliterans shows that bronchiolitis obliterans and tracheobronchomalacia are both caused by a single common process that affects larger and smaller airways in these patients, respectively.

There are also reports of the severe stenosis of the tracheobronchial tree in some cases.

HRCT is routinely used to diagnose tracheobronchomalacia while bronchoscopy is the gold standard for diagnosis [4]. It should be noted that dynamic CT scan images are more sensitive than the conventional static HRCT. About 45–46% of chemical patients have degrees of air trapping, and 15% of them have more than 25% air trapping on pulmonary HRCT. Radiologic studies have shown that air trapping is a common diagnosis on chest HRCT of mustard gas patients 15 years after exposure. On the other hand, air trapping detected on expiratory HRCT is the most sensitive and accurate radiological index of bronchiolitis obliterans. In addition, recent pathological studies have shown that bronchiolitis obliterans is the most important pathologic change in patients with mustard gas–induced respiratory problems [5] (Figs. 4.2 and 4.3).

Assuming that air trapping is a more common finding than tracheobronchomalacia in chemical patients, it can be concluded that air trapping is not a consequence

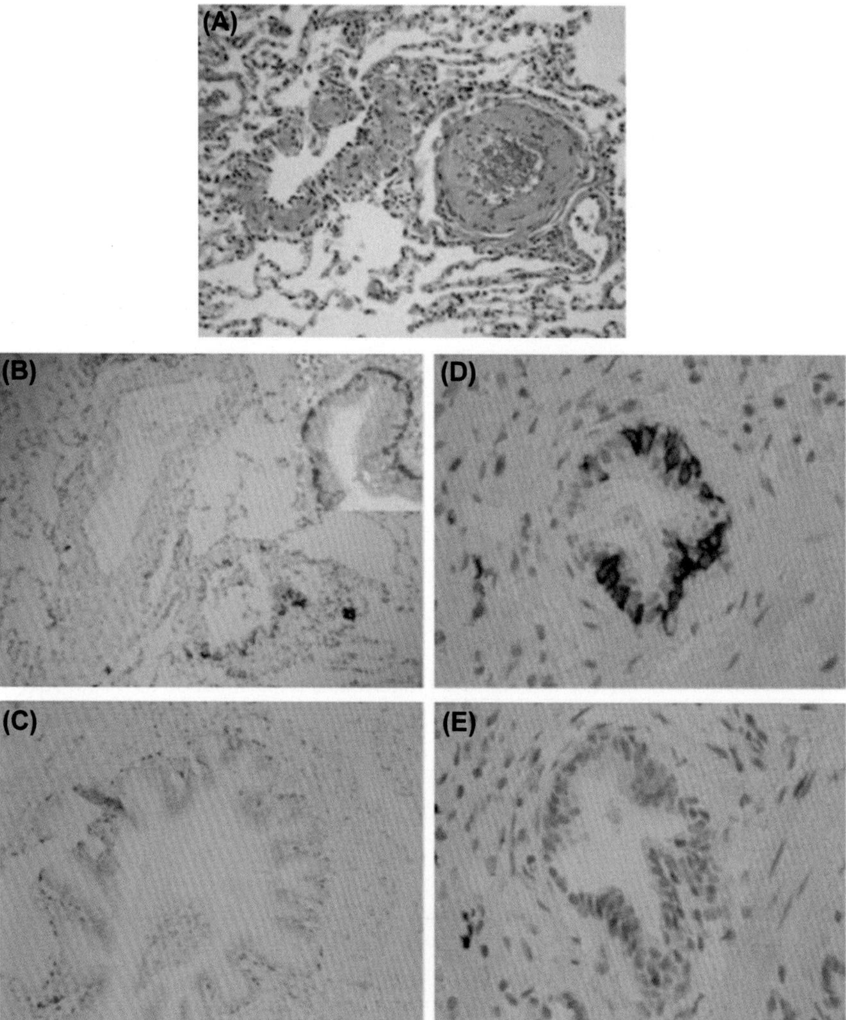

FIGURE 4.2

Pathology specimen of the bronchiole stained with immunohistochemistry techniques in a patient exposed to mustard gas. The evidence of bronchiolitis obliterans is shown in samples stained with H&E (A), anti-CK5 (B), and anti-P63 (C). The last two figures demonstrate samples stained with anti-CK5 (D) and anti-P63 (E) in a patient with hypersensitivity pneumonitis with no exposure to mustard gas for comparison.

of tracheobronchomalacia in this group, and a similar pathological mechanism resulting in small airways disease, ie, bronchiolitis obliterans, may have also resulted in damage to larger airways like tracheobronchomalacia. In other words, mustard gas may affect epithelium in both small and large airways.

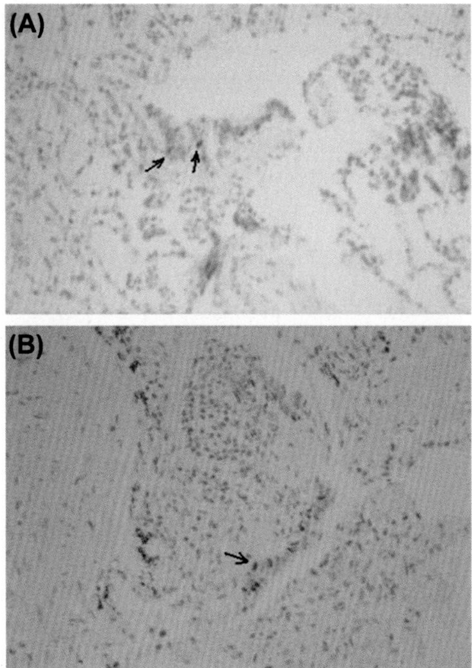

FIGURE 4.3

Evidence of lost bronchioles in the lungs of two chemical patients (A and B). Tissue staining with anti-P63 differentiates the remaining bronchiole tissue from the nonbronchiole tissue.

Both bronchiolitis obliterans and tracheobronchomalacia are known as post-pulmonary transplantation complications. The role of surgical interventions in the development of tracheobronchomalacia after lung transplantation cannot be ignored. Evaluation of respiratory problems in posttransplantation patients suggests a similar pathology resulting in tracheobronchomalacia and bronchiolitis obliterans.

BASICS OF PULMONARY FUNCTION TESTS

The pulmonary function tests evaluated the following four indexes:

1. Air flow (spirometry)
2. Lung volumes
3. Gas exchange (diffusing capacity)
4. Mechanics of the lungs

Accurate interpretation of pulmonary function tests requires definite criteria. The variables that affect the value of these criteria include age, sex, height, race, and hemoglobin concentration. Standard deviation of these variables and their daily variations should be considered while interpreting the results.

Pulmonary function testing includes four volumes and three capacities. Each capacity contains one or more volumes. All volumes, except for residual volume, can be directly measured by spirometry. Residual volume is indirectly measured and is the difference between expiratory reserve volume (ERV) and forced residual capacity (FRC).

The ERV is calculated as below:

1. Calculation of the dilution of an insoluble gas (helium or nitrogen) in a closed environment in which the patient breathes.
2. The gas volume–pressure relationship measured in whole body plethysmography.

Although the gas dilution technique is easy and fast, its accuracy depends on the uniform distribution of the gas in the lung, which does not occur in patients with severe obstructive disease. Therefore, with this technique, lung volumes are underestimated in severe obstructive lung disease. Whole body plethysmography provides more accurate measurements of lung volumes in obstructive lung diseases. Another advantage of this method is that it also measures resistance. However, this technique is difficult and time consuming. Measurement of the airflow through spirometry is the most common test in the evaluation of pulmonary function due to its importance in the detection and treatment of obstructive lung diseases. The patient breathes into a device that measures inspiratory and expiratory flow through making changes in lung volumes and presents the information as flow–volume loop. The pattern of this loop depends on the pulmonary functional disturbance. The loop is used to detect and treat obstructive lung diseases. Forced expiratory volume in 1 second (FEV1)/forced vital capacity (FVC) is the most useful way of showing airway obstruction. In patients with obstructive lung diseases, spirometry is usually performed before and after administering a bronchodilator to evaluate the reversibility of airflow obstruction. In addition, the test might be used as a guideline for treatment. Table 4.1 presents the classification of the patients with respiratory problems. The values stated in the table are based on airflow limitation and disease severity according to spirometry results and clinical findings.

Table 4.1 Classification of the Severity of Pulmonary Disturbance in Patients With Respiratory Problems According to Spirometry Results

Category	Spirometry	Severity (percentage)
Inactive pulmonary disease	$FVC > 80/FEV_1 > 80$	0
Mild	$65 < FVC < 80$ $65 < FEV_1 < 80$	5–20
Moderate	$50 < FVC < 65$ $50 < FEV_1 < 65$	25–45
Severe	$40 < FVC < 50$ $40 < FEV_1 < 50$	50–70

Based on a national agreement, spirometric criteria are used to determine severity of lung dysfunction as percentage, as demonstrated in Table 4.1. These criteria are used in relevant medical commissions.

GAS EXCHANGE CAPACITY OF THE LUNG

The lung gas exchange capacity is determined through the measurement of the diffusing capacity for carbon monoxide. First, the patient inhales a gas mixture containing a low concentration of carbon monoxide and then holds his breath for a certain period of time. Then, the concentration of carbon monoxide in the exhaled air is measured. The difference in the amount of gas is recognized as the gas diffused from the alveolar-capillary membrane and indicates the pulmonary diffusion capacity. Carbon monoxide diffusion capacity is a very good criterion of the absorption and transfer of carbon monoxide and therefore oxygen. However, its abnormal values do not indicate the nature of the gas exchange disorder. Since carbon monoxide mounts hemoglobin, it is important to measure hemoglobin when interpreting the results of this test. Interstitial lung disease, emphysema, pulmonary fibrosis, and pneumonia are among the causes of decreased diffusing capacity of the lung for carbon monoxide (DLCO). Maximum inhalation or exhalation pressure are measured to evaluate respiratory muscles. Several studies in chemical patients have shown that the diffusing capacity of the lung for carbon monoxide does not decrease and is even above normal; therefore, it has no value in the diagnosis and treatment of these patients.

BRONCHIAL CHALLENGE TESTING

When reactive airway diseases (asthma) are suspected, the diagnosis can be confirmed through the measurement of pulmonary volumes and airflow limitation after inhalation of an aerosol containing a potential stimulant. This substance may be a selective bronchoconstrictor like methacholine or nonselective agents to which the patients is sensitive like cold air or organic or nonorganic substances. Repeating this test after treatment with an inhaled bronchodilator demonstrates the reversibility of the obstruction.

CARDIOPULMONARY EXERCISE TESTING

During cardiopulmonary exercise testing, pulmonary ventilation per minute, respiratory rate, expiratory carbon dioxide, heart rate, electrocardiography, and pulse oximetry or arterial blood gases are evaluated. These values are used to determine the share of the heart and lungs in exercise limitations and dyspnea, and the results can help to differentiate physiologic cardiac and pulmonary limitations from other conditions. This test can be beneficial when different findings in a chemical patient are normal but the patient complains of dyspnea.

POLYSOMNOGRAPHY

Polysomnography is used to evaluate patients with sleep disorders. When the patient is sleeping, long-term monitoring of brain activity, electrocardiography, chest movements, diaphragmatic electromyography, pulse oximetry, and recorded video observations are used to make a diagnosis. Using these values, it is possible to confirm a respiratory disorder during sleep and differentiate central from peripheral nervous system disorders.

In a study on 30 chemical patients, the Global Initiative for Chronic Obstructive Lung Disease (GOLD) criteria and polysomnography were used to evaluate the relationship between the severity of lung disease and sleep pattern. The results showed that patients with less severe pulmonary symptoms had more hypopnea and more episodes of REM sleep. Moreover, sleep stage 1 in patients with higher FEV1 and stage 4 in patients with a higher DLCO were dominant sleep stages [6]. These findings indicate better adaption of these patients in more severe stages of the lung disease and the associated dyspnea.

PULMONARY GAS EXCHANGE

Measurement of the arterial blood gases forms the basis of gas exchange evaluation. Arterial blood samples are used to measure pH and partial pressure of oxygen (PO_2). The percentage of oxygen saturation of hemoglobin is naturally calculated with the use of PaO_2, which is an accurate estimate of PaO_2 except for the cases of carbon monoxide poisoning. If there is carbon monoxide poisoning, hemoglobin saturation should be directly measured. $PaCO_2$ and PaO_2 show ventilation and the quality of gas exchange, respectively. The alveolar-arterial oxygen difference (A–a gradient) is calculated to evaluate the decrease in gas exchange and the severity of injury. To calculate the A–a gradient, when the patient is breathing in the room air, measured PaO_2 is subtracted from the calculated PaO_2. The normal value of the gradient is 10 mmHg in young healthy individuals and increases with age and increased percent of inspired O_2. An abnormal A–a gradient is observed in patients with lung parenchymal disease, congestive heart failure, and pulmonary vascular disease. These observations have made the test sensitive but nonspecific. To calculate the acid–base status, pH and PCO_2 of the arterial blood and serum bicarbonate (HCO_3) are used. For this purpose, the Henderson–Hasselbalch equation is employed as the following:

$$pH = pK_a - \log[HCO_3^-]/0.03\ PCO_2$$

where, $pK_a - \log[HCO_3^-]$ represents negative logarithm of the acid dissociation constant.

Table 4.2 presents equations that help to interpret the acid–base status. After detecting an acid–base disturbance, it is important to determine whether the primary cause is respiratory or metabolic. Then, the degree of compensation is evaluated.

Table 4.2 Normal Values of Arterial Blood Gases

$PaO_2 = 104 - (0.27 \times \text{Patient Age})$
$PaCO_2$: 36–44
pH = 7.35–7.45
A–a gradient = 2.5 + 0.21 × age in years

In hypoxic patients, measurement of PaO_2 in the room air and its comparison with situations in which the patient breathes 100% oxygen make it possible to differentiate a pulmonary shunt from a ventilation–perfusion mismatch. In patients with a ventilation–perfusion mismatch, with the increase in inhaled oxygen, PaO_2 increases markedly, while in patients with a shunt, since the blood does not come in contact with the ventilated alveoli, providing supplementary oxygen has no considerable effect on PaO_2. In fact, many shunts have a very low ventilation to perfusion ratio (in other words zero), and most severely hypoxic patients have components of both disorders. If the share of the pulmonary shunt is more, it will not be easy to achieve a desirable PaO_2 through oxygen administration. Increased $PaCO_2$ (hypercapnea) indicates an abnormally low alveolar ventilation, which may be due to little air movement or increased dead space. Sedatives, hypnotics, and narcotics, and central nervous system diseases are among the primary causes of hypoventilation. In normal individuals, hypercapnea resulting from increased dead space can be resolved with hyperventilation within minutes, while in patients with obstructive lung disease or respiratory muscle weakness, ventilation is limited before $PaCO_2$ reaches normal levels. Acute hypercapnea inhibits the central nervous system, which is known as CO_2 narcosis. This disorder can in turn suppress ventilation even more. Mechanical ventilation is usually required for patients with acute $PaCO_2$ increase and the consequent acidosis.

In chemical patients, the resulting hypoxia is immediately resolved with oxygen administration; therefore, no pulmonary shunting exists. In addition, increased $PaCO_2$ is not observed except in very advanced pulmonary disease. For this reason, hypoventilation is not the case. The most common disorder is a ventilation–perfusion mismatch.

RESULTS OF PULMONARY FUNCTION TESTING IN CHEMICAL PATIENTS

In this part, we discuss the importance of pulmonary function tests in chemical patients. Pulmonary function tests can be used to determine the obstructive or restrictive nature of the pulmonary involvement. Moreover, pulmonary function tests are used for the evaluation of the amount and severity of pulmonary involvement and response to treatment. Although the results of different studies can be used to reach a general understanding of the pulmonary function, the results of pulmonary function tests are different in symptomatic and asymptomatic patients. The results of some studies in this area are presented in the following sections.

PULMONARY FUNCTION TESTING IN SUBCLINICAL CONTACT WITH MUSTARD GAS

A study was conducted on 77 persons who were present in mustard gas–contaminated regions for at least one week. The subjects had no symptoms of exposure to mustard gas at that time. The results of the study showed that although these individuals had no clinical symptoms in the acute phase, they developed the complications of exposure in the future. The pattern of pulmonary function testing (PFT) was restrictive in 5%, obstructive in 5%, restrictive and obstructive in 8.82%, and normal in 85.3% [7].

PULMONARY FUNCTION TESTING IN SYMPTOMATIC CHEMICAL PATIENTS

Based on PFT findings in the acute phase, 53%, 1.5%, 18.7%, and 21.8% of the mustard gas patients who did not use appropriate protective gear during exposure had obstructive, restrictive, mixed, and normal PFT patterns, respectively [8].

In another study, the pulmonary function of 130 chemical patients who were hospitalized in Isfahan hospitals, Iran, in the acute phase of injury, was evaluated. The PFT pattern was restrictive in 11.5% and obstructive in 32.3% of the patients. Moreover, 21.5% showed decreased FEV1 and forced midexpiratory flow (FMF) indicating small and large airway obstruction, and 21.5% showed decreased maximal midexpiratory flow (MMEF) indicating small or peripheral airway obstruction. In addition, 14% had a mixed pattern and 41% had normal findings. Considering the above-mentioned results, it could be concluded that an obstructive lesion is the most common abnormal spirometric finding in the acute phase of exposure to mustard gas. Furthermore, near-normal spirometry results comprised a considerable proportion of the findings. It was noteworthy that in two-thirds of the patients older than 41 years, four had a mixed pattern, indicating that the severity of pulmonary involvement increases with ageing [9].

PFT of 35 mustard gas patients with severe pulmonary involvement who were exposed to mustard gas between 6 weeks and 1 year prior to the study revealed that 55% of the patients had an obstructive pattern, 24% had a restrictive pattern, 13% had involvement of small airways, and 6% had a normal pattern [10].

According to a study, the most common spirometric changes and clinical signs and symptoms of the flow–volume loop in patients exposed to sulfur mustard gas were changes in the inspiratory arm of the loop and flattening of the midpart of the inspiratory loop in 18 patients (30%). The most common change in the expiratory arm was a limitation in end-expiratory flow volume in 13 patients (21%). In cases with mild and moderate obstruction, the changes of the loop appeared before spirometric changes, and the changes were only limited to the flow-volume loop. Maximum changes of the flow–volume loop were observed in the fourth month after the injury, while maximum spirometric changes were noted 14 months after the injury.

Moreover, it was reported that distance to the site of chemical attack up to 500 m did not affect the development of the complications but markedly influenced their severity [11].

This study showed that all obstructive disorders in the flow–volume loop were of lower airway obstruction type, and no considerable flow limitation was observed in upper airways in the 2-year study period. In a cross-sectional study, long-term pulmonary complications of 197 chemical patients were evaluated after 10 years, and the results were compared to the results of 86 soldiers with no injury as the control group. In the study, the patients were divided to three groups of asymptomatic, chronic bronchitis, and pulmonary fibrosis, and the results of PFT were evaluated in each group (Tables 4.3 and 4.4). In patients with idiopathic pulmonary fibrosis (IPF), the relationship between the degree of fibrosis and DLCO and PFT findings was investigated, which showed a direct relationship between the severity of fibrosis and DLCO percentage [12]. Since complementary studies ruled out pulmonary fibrosis and interstitial lung disease, the results of this study cannot be used.

The results of a study on 43 mustard gas patients (time between injury and study: 4–7 years) who had long-term signs and symptoms of mustard gas intoxication showed that 53%, 43%, and 5% of the patients had an obstructive, normal, and restrictive PFT pattern, respectively. In fact, the most common pattern of pulmonary function in these patients was the obstructive pattern. On the other hand, many of these patients, despite a normal chest X-ray, had abnormal spirometric and clinical findings. Since the patients who were evaluated at a shorter interval after intoxication had fewer abnormal patterns, their lung injury might be preventable in the course of time [13]. Figs. 4.4–4.6 present spirometric findings of the pulmonary involvement patterns in patients with mild, moderate, and severe lung injury.

In another study that was performed on 407 chemical patients, spirometric changes after 10 years were compared against baseline spirometric findings. According to the results, all the indexes of pulmonary function decreased markedly when compared to the primary evaluation [14].

In a prospective study, Heidarnejad et al. evaluated spirometric parameters (FVC, FEV_1, and PEF) and clinical signs and symptoms of 1872 chemical patients who attended a pulmonary clinic during 10 years. Although the clinical course was rather stable, PEF and FEV_1 underwent statistically considerable changes. Considering the role of ageing in decreasing lung volumes, FEV_1 in these patients decreased 50 ml per year on average. Moreover, because 24% of the patients in the first year were new smokers, no difference in PEF and FEV_1 was observed between smokers and non-smokers in the first year. On the other hand, since these individuals stopped smoking in the following years, the decrease in FEV_1 was stated to be related to the effects of the mustard gas. During the 10 years, PEF parameters had an increasing trend. Because most of these patients were older than 29 years, this finding was assumed to be normal [15]. It should be noted that in this study, inclusion and exclusion criteria were not defined and no information was presented to confirm their contact with mustard gas.

Table 4.3 PFT Results of Patients 10 Years After Mustard Gas–Induced Injury

Asthma Severity	Age (year)	FEV_1 (Lit)		FVC(Lit)		PEF Rate(lit/S)	
		Observed	% predicted	Observed	% predicted	Observed	% predicted
Mild (6 patients)	27.3±0.5	2.7±0.1	83.5±3.06	3.43±0.1	81.05±2.7	7.05±0.2	85.3±27.8
Moderate to severe (15 patients)	32.1±4.9	2.2±0.2	64.2±7.9	3.06±0.1	70.8±3.03	4.85±0.4	56.7±6.16

Table 4.4 PFT Results of 116 Patients 10 Years After Severe Poisoning With Mustard Gas

Severity of Chronic Bronchitis	Age (year)	FEV_1 (Lit)		FVC(Lit)		PEF Rate (lit/S)	
		Observed	% predicted	Observed	% predicted	Observed	% predicted
Mild (27 patients)	36.51±9.63	2.23±0.14	68.98±5.07	3.51±0.28	70.03±4.49	6±0.81	67.80±8.24
Moderate (53 patients)	21.26±6.04	1.45±0.17	47.46±2.2	2.49±0.28	51.79±2.26	4.57±0.39	49.11±6.61
Severe (36 patients)	35.26±2.99	1.19±0.05	23/01±1/54	2.43±0.28	43.57±2.72	2.40±0.30	30.60±3.71

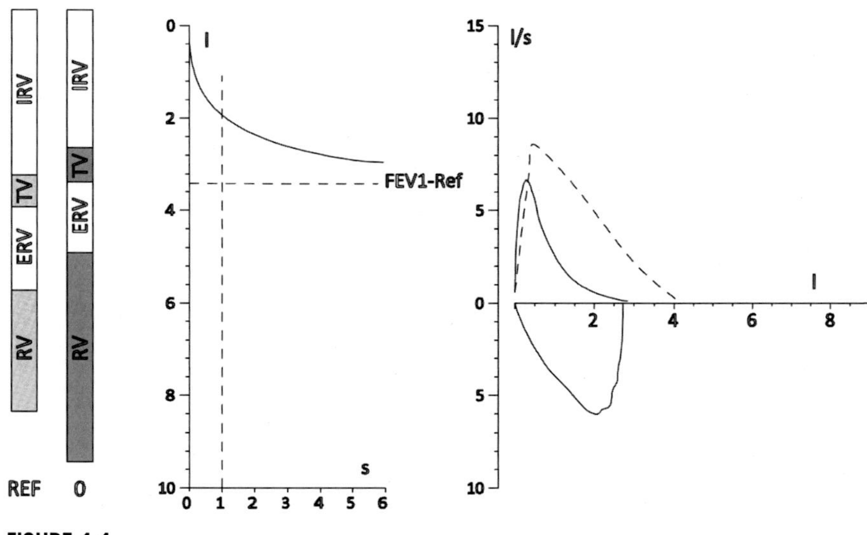

FIGURE 4.4

The flow–volume loop of a 42-year-old male mustard gas patient with a mild obstructive pattern.

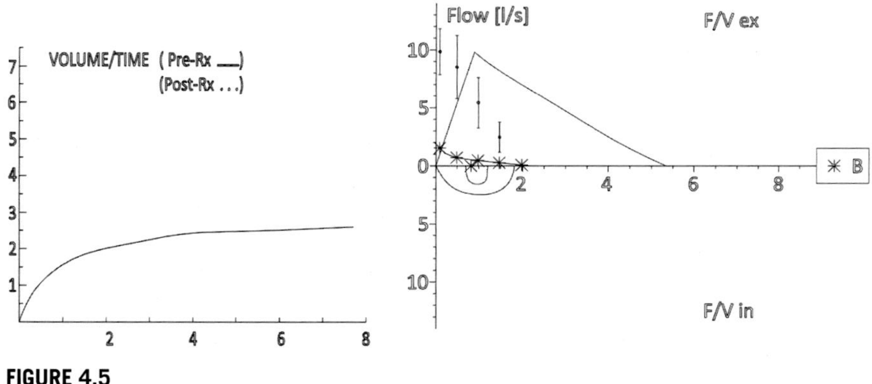

FIGURE 4.5

The flow–volume loop of a 45-year-old male mustard gas patient with a severe obstructive pattern.

Lung capacities of chemical patients were evaluated in similar studies. Evaluation of the different parameters of pulmonary function in 480 chemical patients showed that in different time intervals of 1–150 days after exposure, the disorders in pulmonary spirometric parameters remained unchanged despite treatment. This study suggested that other factors like the distance of the person from the blast area affected the changes of pulmonary volumes and capacities. This study showed a correlation between the distance of the person from the blast area and the decrease in spirometric parameters, but no relationship was found between spirometric values

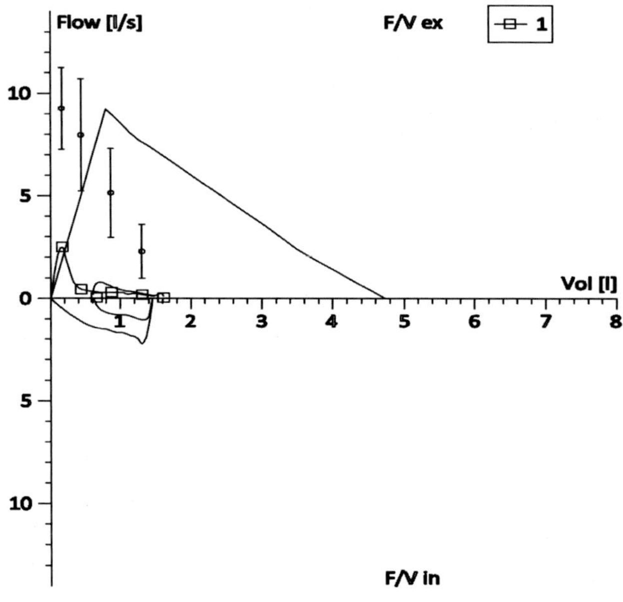

FIGURE 4.6

The flow–volume loop of a 35-year-old male mustard gas patient (70% injury) with a severe obstructive pattern.

and other factors including the time interval between injury and mask use or tobacco use before injury. However, according to the results of this study, if tobacco use was continued after injury, significant decreases in FEV_1/FVC, FEV_1, and FVC were observed. A correlation was also reported between the time of removing the victim from the contaminated area and the variables of FRC, TV, FVC, and FEV_1. Furthermore, the time between injury and the start of treatment and also the time between clothes change and a simple shower were shown to be correlated with FEV_1/FVC, FEV_1, and FVC. With the increase in the time interval between injury and clothes change, shower, and treatment, the mentioned parameters decreased more, and with the use of special clothes (windbreaker jackets) and masks before contamination, the parameters decreased less. Finally, the results of this study demonstrated that a positive allergic history in the patient resulted in increased decrease in lung capacities, especially FMF, FRC, and FEV_1/FVC [16]. Moreover, a direct correlation has been noted between the changes of pulmonary parameters and weight, and an inverse correlation has been detected between age and FMF and FEV_1 [17].

Evidence suggests that the amount and severity of the pulmonary involvement is strongly correlated with the amount of inhaled mustard gas, although this correlation is not always observed; for example, PFT is normal in some patients with a history of symptomatic injury, while some other patients with few symptoms or subclinical injury develop marked disturbances in spirometric parameters in the following years.

This finding suggests the role of genetic differences in the severity of the subsequent pulmonary and skin diseases.

In the early years of the Iraq–Iran war, unreliable devices in terms of validity and accuracy were used for PFT. Many of these devices were not regularly calibrated. Moreover, the criteria proposed by American Thoracic Association were not often regarded in the interpretation of PFT results. As a result, many PFT results in the patients' file lack the required accuracy and validity to determine the type and severity of the lung involvement. Therefore, these test results cannot be used to make accurate reports on the percentage and type of spirometric patterns in these patients. The existing statistical differences are a result of different sampling methods; for example, most of these studies have been performed on individuals with considerable pulmonary involvement, which made them attend treatment centers.

Comparison of the PFT results in recent years with the past shows that the number of patients with an obstructive pattern has increased while the number of patients with a restrictive pattern has decreased. The reason could be little knowledge on the part of the medical staff of the characteristics of a standard spirometry. Moreover, it is possible that the restrictive lesions of the patients transform to obstructive lesions gradually; however, further investigation is required for a definite conclusion. The obstructive pattern is the most common abnormal spirometric pattern in patients with a history of asymptomatic exposure during the war and who developed the signs and symptoms of lung involvement in the following years.

In general, the abnormal PFT pattern in these patients indicates that intoxication with mustard gas results in the airway injury rather than parenchymal involvement. Since pathological and radiological studies also confirm small airways involvement, it is necessary to pay attention to the indexes of involvement in this part and benefit from them for detecting the type of lung involvement.

In general, the near-normal pattern is the most common spirometry pattern in all Iranian chemical patients based on the evaluation of 34,000 patients. Investigations performed in the acute and chronic phases after exposure suggest that the most common abnormal pattern is the obstructive pattern. Other patterns include the mixed and the restrictive patterns. Due to the lack of observing PFT standards in the past, it is difficult to make conclusions based on the changes of spirometric indexes. Decreased pulmonary function and secondary pulmonary diseases (recurrent pneumonia, bronchiectasis, and decreased ciliary function) are expected some years after the injury. However, it seems that exacerbation of the pulmonary problems should be expected in patients with moderate-to-severe involvement of the pulmonary function.

Finally, modern devices like plethysmography and oscillometry can be employed to measure lung capacities and airway involvement more accurately.

The patient's cooperation is not required for performing oscillometry, and its results are very sensitive in determining the pulmonary indexes in patients. If these devices are not available, there should be more emphasis on classic education of the medical staff on the value of spirometry and its standard criteria.

EVALUATION OF THE RELATIONSHIP BETWEEN FATIGUE AND SPIROMETRIC PARAMETERS IN CHEMICAL PATIENTS WITH RESPIRATORY DISORDERS

Fatigue is one of the disabling symptoms in patients with chronic respiratory disorders that extensively affects many aspects of the patient's life. The presence of chronic respiratory disorders in chemical patients was the reason for conducting a study aimed at the assessment of fatigue in chemical patients with respiratory problems and its relationship with the changes of spirometric parameters. In this descriptive cross-sectional study that was performed on 140 chemical patients, after the measurement and registration of spirometric parameters, fatigue was assessed using the translated version of the Multidimensional Fatigue Inventory-20 (MFI-20). The patients were categorized according to spirometric parameters, and fatigue was investigated in different areas according to the severity of pulmonary disorder. The total score of fatigue in chemical patients was 81.6 ± 15.4. The dimension of physical fatigue had the highest (17.3 ± 2.8) score, and the dimension of decreased motivation had the lowest (14.1 ± 4.5) score. According to the Pearson correlation test, there was a significant inverse correlation between spirometric indexes and the score of each dimension of the questionnaire. Chemical patients experience more-than-average fatigue in their everyday life, which has a direct relationship with their spirometric indexes. Therefore, it is important to pay special attention to their fatigue in designing social protection programs and treatment services [18].

CARDIOPULMONARY EXERCISE TESTING IN SYMPTOMATIC CHEMICAL PATIENTS

Cardiopulmonary exercise testing is used for the evaluation of dyspnea in different pulmonary problems [19]. Because many pulmonary diseases mimic the pulmonary disorders of chemical patients, cardiopulmonary exercise testing is beneficial in the diagnosis and differentiation of these diseases and conditions [20]. There are even patients with normal spirometric and HRCT results, which makes the diagnosis of their disorder even more complicated. For this reason, for the first time we used cardiopulmonary testing for the evaluation of the patients exposed to low doses of mustard gas in a case–control study [21]. This study was performed in Tehran, Iran, with a barometric pressure of 670 mmHg. The case group comprised mustard gas patients with exertional dyspnea with normal spirometry and chest X-ray results, unremarkable physical examinations, and no air trapping on HRCT, and the control group included healthy individuals. Cardiopulmonary testing was performed using the Wasserman protocol. In total, 159 patients in the case group were compared with 10 healthy individuals in the control group. The results are presented in Table 4.5. There were significant differences only in WR max, peak VO_2/kg, VO_2 predicted, and RR peak between the two groups.

Table 4.5 The Results of Cardiopulmonary Testing in a Group of Chemical Patients With Respiratory Problems

Variables	Patient Group (Mean ± Standard Deviation)	Control Group (Mean ± Standard Deviation)	*P* Value
VO_2 predicted, L/min (%)	87 ± 12	105 ± 16	<0.001
Anaerobic threshold (%)	30 ± 8	33 ± 8	0.3
Breathing reverse (%)	31 ± 15	26 ± 16	0.39
Ventilation/VCO_2	38 ± 6	34 ± 3	0.14
Ventilation/VEO_2	37 ± 6	37 ± 5	0.96
Saturation baseline (%)	97 ± 1	98 ± 0.7	0.14
Saturation decrease (%)	1.3 ± 1	1.2 ± 0.07	0.9
Ventilation peak predicted (%)	74 ± 15	80 ± 16	0.22
Respiratory frequency peak, Breaths/min	40 ± 8	46 ± 7	0.04
Ventilation threshold at peak (L)	2.0 ± 0.3	2.2 ± 0.5	0.16
Ventilation desaturation/ ventilation threshold at peak	0.16 ± 0.05	0.14 ± 0.05	0.4
End tidal CO_2	32 ± 4	33 ± 4	0.36
End tidal O_2	97 ± 4	95 ± 4	0.3
Maximal heart rate (% of predicted value)	88 ± 7	89 ± 5	0.6
Heart rate recovery	22 ± 20	20 ± 9	0.72
O_2 pulse predicted	101 ± 16	116 ± 9	0.07
Respiratory exchange ratio	1 ± 0.06	1.1 ± 0.07	0.89

The main findings of this study were as follows:

- Venous oxygen pressure is decreased in chemical patients
- The cardiopulmonary response to exercise is not different between the two groups.

The measurement of VO_2 max or VO_2 peak is still the best index for the evaluation of exercise capacity, which should be measured directly because its estimation based on the indexes is not reliable [22]. Decreased peak VO_2 is the starting point for the evaluation of exercise capacity decline.

Considering the previous studies, we tried to prove that many patients, despite the involvement of small airways in laboratory evaluations, would not have significant findings [2].

In fact, this study evaluated the pure effect of mustard gas without considerable pulmonary complications. Finally, it was concluded that although cardiopulmonary exercise testing is recommended for the evaluation of symptomatic patients with normal imaging findings, it is not a useful method for differentiating small airways lesions in patients with pulmonary problems due to exposure to low doses of mustard gas. The Body mass, Obstruction, Dyspnea, and Exercise (BODE) index is a valuable tool for determining the consequences of chronic obstructive pulmonary disease.

A study was conducted with the aim of evaluating the predictive value of BODE in patients with mustard gas–induced pulmonary disease. Eighty-two patients with different severity levels of pulmonary involvement were included in this study. Standard spirometric parameters, pulse oximetry, health-related quality of life, and the BODE index was recorded for all patients. A significant inverse correlation was detected between the predictive value and oxygen saturation. Moreover, a significant relationship was found between the predictive value and the quality of life, while predictive value had no relationship with age and disease duration. The results of the study showed that the predictive value was associated with important clinical parameters and could be used for clinical assessment of the patients [23].

REFERENCES

[1] Ghanei M, Mokhtari M, Mohammad MM, Aslani J. Bronchiolitis obliterans following exposure to sulfur mustard: chest high resolution computed tomography. Eur J Radiol November 2004;52(2):164–9.

[2] Beheshti J, Mark EJ, Akbaei HM, Aslani J, Ghanei M. Mustard lung secrets: long term clinicopathological study following mustard gas exposure. Pathol Res Pract 2006;202(10):739–44.

[3] Veterans at risk: health effects of mustard gas and lewisite. Committee to Survey the Health Effects of Mustard Gas and Lewisite. Washington, DC.: National Academy of Sciences, Institute of Medicine, National Academy Press; 1993.

[4] Ghanei M, Moqadam FA, Mohammad MM, Aslani J. Tracheobronchomalacia and air trapping after mustard gas exposure. Am J Respir Crit Care Med February 1, 2006;173(3):304–9.

[5] Ghanei M, Chilosi M, Mohammad Hosseini Akbari H, Motiei-Langroudi R, Harandi AA, Shamsaei H, et al. Use of immunohistochemistry techniques in patients exposed to sulphur mustard gas. Pathol Res Int 2011;2011:659603.

[6] Vahedi E, Taheri S, Alaedini F, Poursaleh Z, Ameli J, Ghanei M. Correlations of sleep disorders with severity of obstructive airway disease in mustard gas-injured patients. Sleep Breath June 2012;16(2):443–51.

[7] Ghanei M, Fathi H, Mohammad MM, Aslani J, Nematizadeh F. Long-term respiratory disorders of claimers with subclinical exposure to chemical warfare agents. Inhal Toxicol July 2004;16(8):491–5.

[8] Sohrab pour H. Observation and clinical manifestation of patients injured with mustard gas. Med J Islamic Repub Iran 1987;1:32–7.

[9] Golshan M, Mozaffari A. Evaluation of PFT changes in the acute phase of mustard exposure in 130 chemical patients. [MD thesis]. Isfahan University of Medical Sciences; 1988.

[10] Sohrabpour H, Masjedi MR, Bahadori M. Late complications of sulfur mustard in respiratory system. Med J Islamic Repub Iran 1988;2(3):171–3.

[11] Kh A, Salehi R. Changes of the flow-volume loop 2 years after exposure to mustard. In: The first international congress of war gases in Iran. Mashhad. June 1988. Article number: 67.

[12] Emad A, Rezaian GR. Immunoglobulins and cellular constituents of the BAL fluid of patients with sulfur mustard gas-induced pulmonary fibrosis. Chest May 1999;115(5):1346–51.

[13] Keshmiri M, Bijani KH, Bavandi M. Evaluation of the long term effects of war gases on PFT in victims of Iraq-Iran war in the first 6 months of 1992. In: Emam Reza and Ghaem hospitals. J Mashhad Univ Med Sci 36(45):3–7.

[14] Ghanei M, Eshraghi M, Peyman M, Alaeddini F, Jalali A, Sajadi V. Pulmonary function test trend in adult bronchiolitis obliterans. Tanaffos 2007;6(3):40–6.

[15] Heidarnejad H, Zendehdel N, Dastgiri S. Temp oral trend of clinical and spirometric parameters in mustard gas victims: a ten-year study. Arch Irn Med 1998;1(1):13–6.

[16] Eftekhar Hoseini SA, Motamedi F, Semnanian S. Relationship of age, height, weight, cigarette addiction, duration of injury, and type of chemical agent with lung volumes and capacities. In: The first international congress of war gases in Iran. Mashhad. June 1988. Article number:54.

[17] Semnanian S, Eftekhar Hoseini SA, Motamedi F. Analysis of spirometric data of chemical patients and their comparison with normal values. In: The first international congress of war gases in Iran. Mashhad. June 1988. Article number:55.

[18] Najafi Mehri S, Pashandi S, Mahmoodi H, Ebadi A, Ghanei M. Assessment of fatigue and spirometery parameters in chemical war victims with respiratory disease. IJWPH 2010;2(4):29–35.

[19] Wasserman K, Hansen JE, Sue DY, Whipp BJ, Casaburi R. Principles of exercise testing and interpretation: including pathophysiology and clinical applications. 3rd ed. Philadelphia: Lippincott Williams and Williams; 1999.

[20] Chan A, Allen R. Bronchiolitis obliterans: an update. Curr Opin Pulm Med 2004;10:133–41.

[21] Aliannejad R, Saburi A, Ghanei M. Cardiopulmonary exercise test findings in symptomatic mustard gas exposed cases with normal HRCT. Pulm Circ April–June 2013;3(2):414–8.

[22] Gibbons RJ, Balady GJ, Beasley JW, Bricker JT, Duvernoy WF, Froelicher VF, et al. ACC/AHA guidelines for exercise testing. A report of the American College of Cardiology/American heart association task force on practice guidelines (Committee on exercise testing). J Am Coll Cardiol July 1997;30(1):260–311.

[23] Lari S, Ghobadi H, Attaran D, Kazemzadeh A, Mahmoodpour A, Shadkam O, et al. The significance of BODE (BMI, obstruction, dyspnea, exercise) index in patients with mustard lung. J Cardio Thorac Med 2013;1(1):7–11.

CHAPTER

5 Signs and Symptoms of Exposure to Mustard Gas

After acute exposure to mustard gas, the signs and symptoms of the poisoning can be divided into two time phases: the acute phase, which is the time between the exposure until weeks after; and the chronic phase, which continues to months, years, or often to the end of the patient's life. The acute phase is critical and, because of the variety of the signs and symptoms, can be life threatening. The chronic phase usually does not threaten the survival of the patient and mostly affects the patient's quality of life and causes disability. Here, we separately discuss the symptoms and characteristics of these two phases.

SYMPTOMS OF THE ACUTE PHASE

The main feature of mustard gas exposure at low doses is an incubation period, without any clinical signs and symptoms within a few hours after the exposure. However, signs and symptoms develop in coming days and months [1]. Duration of this period depends on the nature of exposure, ambient temperature, and probably the individual's characteristics. Some people obviously show more sensitivity to mustard gas compared to others.

During chemical injury, airways, eyes, and skin are in direct contact with mustard gas and develop the main clinical symptoms. When the lungs or skin absorption rate is high, it can cause complications in the hematopoietic system, gastrointestinal system, and central nervous system. In such cases, the symptoms of mustard gas poisoning will be subject to multiple organ dysfunction syndrome (MODS).

Damage to the respiratory system includes acute edema, inflammation, and the destruction of epithelial tissues of the airways. Depending on dose, the damage may range from mild to severe. Acute destruction includes the damage of the epithelium and subsequent formation of a pseudomembrane, which may block the airways and cause death.

Although the bronchioles are also affected, in most cases the most severe damage is caused in the larynx, the trachea, and bronchi. Edema forms in basement membrane causing the infiltration of blood cells. In some cases, after the exposure to high doses, the damage develops to the deeper alveolar regions causing generalized edema of the lungs. Reaction of increased acute allergic sensitivity to mustard gas has not comprehensively been studied, however, it can be followed in the chronic phase [2].

Mustard Lung. http://dx.doi.org/10.1016/B978-0-12-803952-6.00005-8

In case of MODS, respiratory tract infections and acute bronchopneumonia are common complications after the exposure to mustard gas. It seems that most of deaths after inhalation of the sulfur mustard are due to a decrease in white blood cell counts after the infection of bone marrow and lung infection, and their septicemia [3]. Immune system weakness resulting from the systemic absorption of the mustard gas probably plays an important role in the pathogenesis of this infection.

Runny nose, often severe, is one of the common symptoms after exposure to mustard gas. Bleeding from the nose, because of the severe damage to the mucosa, might be seen in patients who have been severely affected. Inflammation and ulceration of the mouth palate, nasopharynx, oropharynx, and larynx lead to hoarseness, and the person may temporarily lose his or her voice.

Reports from World War I (WWI) show that laryngeal edema or severe spasm can be severe enough to make tracheotomy surgery necessary in order to save the patient's life. Willems's reports on the bronchoscopic findings in some patients showed bilateral erythematous of the respiratory tract mucosa associated with bleeding and pus discharge [4].

Even separated pieces of mucosa due to necrosis have been seen. Coughs may be severe and purulent mucous sputum may form. In extreme cases, in the following days, acute respiratory distress syndrome (ARDS) and pneumonia may develop.

During the WWI the workers in factories producing mustard gas were at risk of infection. Among the workers of a factory in Great Britain, about 1400 employees were injured and every 3 months most of the workers suffered from accidental burns and blisters. (It is thought that some workers were injured more than once.)

Conditions in the main plant in France that produced three-quarters of the mustard gas for allies during WWII were similarly unpleasant. Statistics showed that 90% of the staff lost their voices and 50% of them were constantly coughing.

Also, after long-term exposure to small amounts of gases in the workplace atmosphere, their primary skin defense was weakened. The main consequence was intense itching of the body, making it impossible for them to rest. The workers severely lost their stamina.

Table 5.1 shows the development of expected symptoms and signs after having severe contact with mustard gas. Primary clinical signs of exposure to various doses of mustard gas in different body organs are given in Table 5.2 [5].

Drinking water or food contaminated with sulfur mustard causes different signs and symptoms from what has been previously mentioned. After several hours, symptoms such as nausea, vomiting, and abdominal pain, and in case of severe poisoning, blood vomiting, diarrhea, illness, and toxicity are expected.

During the 8-year war that Iraq imposed on Iran, several studies investigated the patients' conditions exposed to mustard gas. The main and acute pathologic findings reported include demyelization, pulmonary dysfunction, ocular symptoms, conjunctivitis, corneal ulcer, gastrointestinal symptoms (dysplasia, diarrhea, and vomiting), septicemia and liver and kidney failure. In the first 24 h after contact with mustard gas, skin blisters containing a yellowish liquid are the unique sign of acute exposure to mustard gas, making it distinguishable from other chemical injuries.

Table 5.1 Clinical Signs and Symptoms After Severe Exposure to Mustard Gas

Time	Signs and Symptoms
20–60 min	There are often no signs or symptoms. Nausea, vomiting, eye pain (as sharp pain) are sometimes observed.
2–6 h	Nausea, fatigue, headache, eye inflammation, severe eye pain, tearing, blepharospasm (spasm of the eyelids), photophobia, rhinorrhea are observed.
6–24 h	Face and neck rash and sore throat are observed. Hoarseness may develop and the person may lose their voice. Heart rate and respiratory rate increase. Increased severity of the previously mentioned complications, inflammation of the inner side of thighs, genital areas, perineum, buttocks, and armpits followed by development of blisters. Large blisters may hang and become full of clear yellowish liquid. Death within 24 h after exposure is rare.
48 h	General conditions become worse. Blisters get bigger and cough starts. Purulent necrotic mucosa may be seen in the sputum. Severe itching of the skin is common. Skin pigmentation increases.

Skin burns are often in range of grades 1 and 2, and in extreme cases can be grade 3 as well. Drasch et al. reported unmetabolized mustard gas in the tissue biopsies of Iranian patients [6]. The sample tissue was the adipose tissue or any other fat-containing tissue. Mustard gas was found in the fat tissue of thighs, abdominal skin, and subcutaneous adipose tissue. In addition, in the acute phase, a low concentration of unmetabolized mustard gas was found in the lung, spleen, liver, blood, and urine samples.

One of the severe life-threatening effects of mustard gas exposure, which has been proved by different researchers, is the severe dysfunction of the immune system (especially in cases of injuries and blisters) that can create the grounds for opportunistic infections and resistance to common antibiotics. This can lead to septicemia and death (for further explanation, please refer to the physiopathology in Chapter 3). Secondary infection in all the three systems of respiratory, vision, and skin is probable during the early days after exposure.

Eye symptoms include eyelid swelling, irritability, and extreme conjunctivitis, which lead to transient blindness. Due to the cholinergic effects, meiosis may be seen, which should be differed from other toxicities. Direct eye exposure to liquid mustard gas causes more severe and probably permanent eye damage.

LATE SYMPTOMS OF EXPOSURE TO MUSTARD GAS

The sulfur mustard toxic effects can be seen both primarily and with delay in the eyes, skin, and respiratory organs. Skin and eye injuries may remain unchanged for a long time or may decrease. But the respiratory complications are the most common late and delayed complications. These problems can be debilitating and may progress over time [7].

Table 5.2 Primary Clinical Signs of Exposure to Various Doses of Mustard Gas in Different Body Organs

Target Organ	Severity	Symptoms	Incidence Time (Hour)
Eye	Mild	Tearing Itching Burning Foreign body sensation	4–12
	Moderate	Redness Eyelid edema Little pain	3–6
	Severe	Swelling of eyelids Erythema Possible damage to the cornea Severe pain	1–2
Airways	Mild	Rhinorrhea Sneezing Nosebleeds Hoarseness Coughing	6–24
	Severe	The above-mentioned symptoms plus: Productive cough Moderate-to-severe shortness of breath	2–6
Skin	Mild	Erythema	2–24
	Severe	Development of blisters	4–12

The effects of mustard gas are influenced by several factors. Exposure intensity is highly important, as typically, the symptoms at the time of exposure can indicate the severity of poisoning. It should be noted that environmental and genetic factors are among those that can alter the effects of mustard gas.

In a retrospective cohort study of 1337 soldiers exposed to mustard gas, Zarchi et al. defined the relationship between some factors and long-term pulmonary complications. These factors include age, cigarette smoking, frequency of being exposed to gas, and the use of gas masks. Their study showed that the cumulative incidence of complications was 31.6% and it was determined that increase in age significantly increases the risk of pulmonary complications.

Therefore, the risk of complications was 0.75 per 1000 in the first year after the exposure, and 9.76 per 1000 in the seventh year. Their study also showed that the use of masks decreases pulmonary complications and it seems that gas concentration, time of the exposure to the high dose, temperature, and environmental conditions are also influential in the incidence and severity of side effects [8]. Our study showed that long-term effects of mustard gas are not dependent on the exposure dose [9].

Determining the relationship between the intensity of the initial symptoms and the incidence of delayed complications requires a comparison between the complications with the history of the initial exposure in order to determine if the exposure was severe or mild. Determination of the exact dose and intensity of exposure in war conditions, especially when the gas is distributed in an indefinite and unrestricted area, is not possible.

Despite all the evidence on the long-term pulmonary effects in severe exposure to mustard gas, little information has been obtained on mild exposure and its delayed and late effects. So far, there is not sufficient information to prove the theory that intensity of the exposure and initial symptoms can be used for the prognosis of the delayed pulmonary complications.

UPPER AIRWAYS

Akhavan et al. studied the abnormal findings on larynx in 50 cases, 20 years after the acute chemical phase. In this study, dysphonia including harshness in 14% of cases, hoarseness in 32% of cases, and chronic laryngitis was observed in most cases.

According to the study, long-term neurotoxic effects of mustard gas may cause vocal cord paralysis; while synechia and the formation of vocal cord nodules may be caused by infection of the larynx. In addition, hypertrophy of the pseudo vocal cords may be caused due to the inability to correctly use the edematous vocal cords, and dysphonia may occur as a result. All these signs confirm the existence of chronic laryngitis [10].

Tracheobronchomalacia develops in some victims injured by mustard gas after 15 years. In a study using bronchoscopy on patients with chronic cough, we reported tracheobronchomalacia in 14% of the cases. Tracheobronchomalacia is usually caused due to weakness in the cartilage of the walls of the airway and its supporting areas and results in severe collapse of the central airways.

Tracheobronchomalacia has primary and secondary types. The primary type causes infection and stridor in children's lungs. Bronchial diseases form in the secondary type. Extensive panbronchiolitis may be involved in intensification of tracheobronchomalacia [11].

Sinusitis is one of the most common causes of chronic cough and is the most common symptom among the patients exposed to sulfur mustard. In a case–control study among patients to evaluate chronic persistent cough, paranasal sinus computed tomography (CT) scan was performed in 39 chemical victims and 35 patients without a history of exposure to sulfur mustard with continuous chronic coughs.

Except for one case, all the CT scans showed some sinus complications. Mucosal abnormalities were observed in 30 (76.9%) cases of chemical victims. The abnormality was severe in eight patients, but no significant difference was observed between the results of the two groups. Thus, various types of sinus abnormalities were common in both groups of chemical and nonchemical victims, which did not follow any particular pattern. It was concluded that using the current methods to investigate the causes of chronic cough can lead to a delay in diagnosis [12].

LOWER AIRWAYS

Delayed pulmonary disorders are the most common complications following injury by mustard gas. The prevalence of these complications in the first year has the lowest rate, but increases gradually. The main pulmonary complications of mustard gas victims include: bronchiolitis obliterans, bronchiectasis caused by chronic bronchitis, asthma-like symptoms, and constriction of the main airways. The direct effects of mustard gas particles, in sizes of 1–5 μm, occur when they reach the small bronchioles. Other effects on the respiratory airways are due to the systemic absorption of sulfur mustard.

Several studies have pointed out that the intensity of primary symptoms caused by mustard gas exposure is associated with risk of developing obstructive lung diseases. In comparison to patients with moderate-to-severe symptoms, the risk of developing obstructive disease in patients with mild early symptoms is less. Functional pulmonary abnormalities detected on pulmonary function tests (PFTs) are the most common pathological findings among pulmonary diseases after exposure to mustard gas, which are not associated with moderate to severe primary symptoms of the patients.

It has been reported that sometimes following initial exposure, in patients with mild-to-moderate symptoms the pulmonary function test remained normal in the chronic phase and fewer of them developed delayed obstructive pulmonary symptoms. Evidence showed that if the exposure to mustard gas is so intense that it leads to hospitalization of the patient in the acute phase, the probability of the incidence of delayed pulmonary complications will also increase.

However, the alteration of symptoms from moderate to severe or frequent periods of hospitalization after the exposure did not change the results. So, it seems that other factors such as individual susceptibility in patients to the intensity of primary symptoms or hospitalization were of greater value [13,14].

Injuries caused by the exposure during the war are not simple in nature. Several factors affect the severity of injuries resulting from the exposure to mustard gas. Air temperature and humidity, the moisture level and the area of the exposed skin, personal protective equipment, the speed of wind blow and its direction, the level of activity at the time of exposure, and the susceptibility of the individual are all factors that can alter the effects of mustard gas [15].

Effective exposure, which is able to cause premature symptoms, is essential for the development of late pulmonary complications. But, in case of the development of early symptoms, their intensity will not indicate the intensity of late pulmonary complications. Also, the chest high-resolution computed tomography (HRCT) findings, as the diagnostic method of choice in the chronic phase, are not associated with early pulmonary symptoms [16]. Our findings suggest that increase in the intensity of early pulmonary symptoms does not cause progressive findings like air trapping or mosaic diffusion. These findings were repeatedly observed in radiographic images of exposed symptomatic and asymptomatic individuals.

As mentioned earlier, histopathologic findings and radiology findings (HRCT) clearly show the diagnosis of bronchiolitis obliterans in chemically injured victims [16]. In people exposed to mustard gas, bronchiolitis must be considered as a major long-term complication.

The findings of this study are not comprehensive enough to determine the mechanism of the development of bronchiolitis; thus, it can only be said that bronchiolitis obliterans is not associated with the intensity of symptoms or the duration of hospitalization after the exposure, and to explain the pathogenesis of the disease, other factors must be considered.

While the intensity of early symptoms and the dose of mustard gas cannot predict the prognosis of the patient, the individual's susceptibility and immune factors such as TGF-β have greater certainty in determining the development of late pulmonary complications [17,18].

CHRONIC COUGH

Chronic cough is one of the most common causes of outpatient visits at the community level. The prevalence of chronic cough is associated with smoking status. Smoking is the cause of 5–40% of chronic bronchitis and is the most common cause of chronic cough in the general population.

In adults with normal immune systems in the absence of angiotensin converting enzyme (ACE) inhibitor usage, postnasal discharge (PND), asthma, and gastroesophageal reflux are the reasons for 90–100% of chronic coughs. Although chemical victims often suffer from bronchiolitis obliterans, our study showed that among the reasons mentioned, gastroesophageal reflux and then bronchiolitis play the most important roles in the incidence of cough and exacerbation of chronic cough among chemically injured patients [12,19].

Bronchial spasm caused by reversible smooth muscle contraction has been observed in most patients with bronchial asthma. It has also been observed in more than half of the patients with exogenous allergic alveolitis, in more than one-third of patients with chronic bronchitis and pulmonary tuberculosis, and in one-fifth of patients with pulmonary sarcoidosis.

In addition, it has been proven that bronchial spasm plays an important role in the development of bronchoconstriction in patients with chronic bronchitis. Studies show that in 66% of chemically injured victims, the chronic coughs are caused by bronchoconstriction. Hence, it is important to study bronchoconstriction first, and in case of approval, start its treatment [19].

Gastroesophageal reflux disease (GERD) is another cause of chronic cough. Lack of information on this can put the patient in a vicious cycle. Because, GERD can stimulate the cough, inadequate treatment of reflux may make the cough–reflux cycle become permanent. Our studies showed that pepsin and trypsin exist in the lavage fluid of all the chemically injured victims exposed to mustard gas, and bile acids are found in the lavage fluid in 86% of the cases [20]. Also, pepsin concentration in a case group was more than that in the normal subjects in the control group [21].

A study by Irwin et al. showed that antireflux treatment with H_2 receptor antagonists or prokinetic drugs can treat coughs in 70–100% of the adult patients. Because of the high prevalence of GERD in chemically injured victims, the mentioned treatments are recommended in all patients with chronic cough, even in those without distinct GERD symptoms [22].

According to some reports, for the chronic cough, there were two reasons in 18–62% of patients referred to medical clinics, and three reasons in more than 42% of patients. Being aware of the fact that chronic cough can be caused by more than one factor at the same time, is necessary for an accurate diagnosis and successful treatment of the cough.

The importance of this increases especially when the results of the study of cases exposed to mustard gas are compared to those who were not exposed. Also, more than 90% of the patients in our study had a combination of causes for chronic cough. Thus, we can conclude that despite the known causes of chronic cough, such as chronic bronchitis, in patients exposed to mustard gas, further assessment should be done on patients with chronic bronchitis caused by mustard gas, in particular, to investigate other causes of chronic cough, and especially in the uncontrolled chronic coughs or recently intensified cases.

Moreover, since the number of factors involved in patients exposed to mustard gas is considerably higher than in those who are not chemically injured, we offer an evaluation of possible exposure to chemical gases in patients with chronic cough. This means that potential exposure to chemical gases should be considered as a cause of chronic cough. As mentioned, these findings can be helpful in the diagnosis of clinical exposure to mustard gas especially in victims with a history of participation in the war-affected fields, which may have been infected by delayed progressive complications.

It should be emphasized that chronic cough in the chemically injured victims should not be considered only as a result of the exposure to mustard gas; other causes of chronic cough should also be examined. Surely, coughs cannot be treated appropriately unless a proper diagnosis has been made. When a treatment for a diagnosis improves the cough in the short term, the diagnosis should be considered as one of the causes for the cough and should be continued until other diagnoses are made.

PREVALENCE AND DIAGNOSIS OF EMPHYSEMA IN CHEMICAL PATIENTS

Emphysema is morphologically defined as persistent distal airway dilatation to the terminal bronchioles and the damage of their walls without obvious fibrosis. Aging and cigarette smoking are the main causes. Alpha-1 antitrypsin enzyme deficiency is a genetic factor that can lead to early onset of emphysema.

The old methods for identifying emphysema are PFTs and chest X-rays. Today, these methods are not sensitive enough for the early diagnosis of functional and morphological abnormalities of alveoli and their connected airways. The use of HRCT is recommended for accurate assessment of the extent and intensity of emphysema alterations.

In a cross-sectional study on 20 symptomatic smokers with a history of mild exposure to mustard gas and 20 smokers without the exposure, function tests and chest HRCTs were done for all patients to detect emphysema. Sensitivity, specificity, and negative predictive values of respiratory function tests were measured.

In the group with mild exposure to mustard gas, spirometry did not have the ability to detect emphysema, while chest HRCT detected five patients. In the smoker group more cases of emphysema (11 out of 20—55%) were observed, compared to the group with mild exposure to mustard gas (5 out of 20—20%). No case of alpha-1 antitrypsin deficiency was detected in the patients [23].

It can be concluded that in cases with a history of exposure to mustard gas, cigarette smoking can cause emphysema at an earlier age. In this group, HRCT is more useful for early diagnosis than the pulmonary function tests because the pulmonary function tests were normal in this group.

In our study, no connection was found between air trapping in HRCT and pulmonary PFTs. A PFT may not be able to accurately detect the longitudinal changes in acinar structures and the changes that are caused by specific factors like age and cigarette smoking. In cases where an additional pulmonary risk factor like exposure to toxic gases exists, the symptoms extend and the diagnostic approach may differ from other patients. Given the high prevalence of such exposure in cities and industries, it seems necessary to study emphysema in patients with more than one risk factor.

Bronchiectasis and bronchial wall thickening can be clearly seen in HRCT images. Some studies have suggested that sulfur mustard, combined with other known side effects, can lead to emphysema. But recently, after removal of confounding factors like cigarette smoking, it was demonstrated that emphysema is not associated with the exposure to sulfur mustard and is not one of its late complications.

Hence, the observation of emphysema in the lungs of a chemically injured victim confirms that, apart from the exposure to mustard gas, other factors like smoking had a role in its incidence. In chest HRCT, air trapping and mosaic pattern are the most common radiological findings and have been reported both in symptomatic and asymptomatic patients. The group with mild exposure to mustard gas showed a higher degree of air trapping compared to the smoking group.

In general, chemically injured patients had a higher rate of air trapping in comparison to those without exposure. Cough and sputum had a higher rate among smokers with a history of exposure to mustard gas. However, other signs and symptoms did not differ between the groups. Central and paraseptal emphysema were more prevalent in the smokers group.

In our study, although PFT (FEV_1/FVC) in the smokers with no history of exposure to mustard gas was an appropriate diagnostic criterion, it was not a suitable criterion for patients with emphysema who had a history of exposure to mustard gas.

Smokers with a history of exposure to mustard gas showed earlier pulmonary symptoms than other smokers. They were, on average, 12 years younger than the others.

Thus, except for alpha-1 antitrypsin deficiency, other factors may lead to symptomatic emphysema at younger ages than is expected. Compared to nonexposed smokers, exposed smokers suffer similar symptoms of emphysema at a lower rate and at younger ages. Smokers of the exposed group had more risk factors for respiratory complications. They had more coughs and sputum, even in younger ages, and showed a higher degree of air trapping.

We can conclude that in some circumstances, with previous exposure to chemical agents like mustard gas, emphysema may develop at an earlier age than expected. Given that study patients were those with mild exposure to mustard gas, it can be concluded that even mild exposure to toxic gases along with smoking may increase the risk of developing emphysema and is associated with respiratory symptoms.

So, even in patients who do not remember any acute respiratory exposure to toxic gases, in time of early development of emphysema, prior exposure to toxic gases should be considered as a differential diagnosis. In other words, in such circumstances, spirometry does not have the necessary accuracy, and chest HRCT should be the diagnostic method of choice for the evaluation of emphysema [24].

In previous studies on patients with symptoms of emphysema, a considerable relationship was found between chest HRCT and lung function tests. But, recent studies have reported that HRCT in smokers can detect emphysema changes despite normal PFT. Sanders et al. found evidence of emphysema in 69% of patients who did not demonstrate functional emphysema [25]. But, the results of pulmonary function tests in all the participants, regardless of air trapping, was reported to be in the normal range. FEV_1 and $FEV_1/FVC\%$ can also be close, or almost equal, to normal ranges in smokers with or without the emphysematous lesions.

The apparent difference may be due to the detection of local abnormalities in CT scans. In comparison, spirometric measures report more generalized measures of lung function. Thus, smokers with a history of prior exposure to respiratory toxins are susceptible to develop emphysema at younger ages, and spirometry does not have the ability to detect emphysema in them. Thus, the routine screening tests in patients with a known history of exposure to toxic gases shall include chest HRCT and pulmonary function tests, particularly in patients who are symptomatic.

Although the residual volume, total lung capacity and inspiratory capacity, may be used for predicting pulmonary tract obstruction, FEV_1/FVC (usually performed in routine spirometry) would not be helpful. It has been previously shown that chest HRCT may be more useful in the early diagnosis of lung diseases associated with smoking. HRCT can identify emphysema in symptomatic smokers even with normal respiratory function tests. Chest HRCT sensitivity is 100% and its specificity is 91% in identifying the pathology [26]. So, the chest HRCT should be undertaken for every smoker with a mild history of exposure to respiratory toxins.

It should be noted that exposure to mustard gas by itself does not cause emphysema. As in the normal population, chest HRCT in patients with exposure showed lower areas of emphysema or air trapping in nonsmokers in comparison to smokers.

Emphysema was reported to be in association with mustard gas exposure. However, considering smoking effect as a confiding factor, the evidence did not confirm the relationship between emphysema and exposure to mustard gas. No case of emphysema was reported in nonsmokers because smoking was the main reason of emphysema in individuals exposed to the mustard gas and in the general population.

As noted, emphysema increases markedly with age. The lung undergoes morphological changes with age including: increased air in airways and alveoli, reduced complexity of the alveolar surface or surface-to-volume ratio, the loss of alveolar wall tissue, elastic tissue, and bronchial muscles, and increase in emphysema.

Considering the fact that in the mentioned study the cases in the smoking group were older than the group exposed to low amounts of mustard gas, it can be concluded that the use of chest HRCT will be more useful in the group exposed to low amounts of mustard gas because they are in earlier stages of the disease. The emphysema rate was higher in the smokers group compared to the chemically injured victims. Emphysema and chronic bronchitis impose a heavy burden on society. Except for the costs of illness, the intensity of illness is higher in patients with emphysema.

This study may help us in the diagnosis of emphysema at younger ages and in early stages in smokers with a history of exposure to respiratory toxic substances and enable us to reduce the hospitalization rate and decrease health care costs.

Previous studies showed that awareness of the risk factors (eg, alpha-1 antitrypsin deficiency) may lead to positive changes in health, such as trying to quit smoking. Therefore, smokers who have a history of exposure to toxic gases must be informed of emphysema risk. The warning may encourage them to quit smoking.

This study showed the value of chest HRCT in current smokers with emphysema symptoms. Even in nonsymptomatic individuals, a noticeable air trapping is probably pathologic. Considering that the possibility of asthma has been studied and rejected in these patients, the air trapping may be a result of inflammatory bronchiolar damage. Future studies on asymptomatic patients will help us to diagnose the emphysema in its early phases.

In summary, in the smokers with additional risk factors, such as exposure to respiratory toxic gases, emphysema begins at an earlier age. In this stage, the spirometry is still normal.

Chest HRCT should be considered as an appropriate tool for early diagnosis of emphysema in smokers with a history of exposure to toxic gases. This additional risk factor can exacerbate the symptoms in early stages. Given the mentioned evidence, it seems that the early diagnosis of emphysema can be made before the onset of symptoms.

CARCINOGENIC PROPERTIES OF MUSTARD GAS

Although the carcinogenic property of the mustard gas from repeated exposure has been proven, like in factory workers in which toxins are produced, the ability of mustard gas to cause cancer in acute exposure cases and from a single-dose exposure is still disputed.

In the beginning, there was no evidence of a carcinogenic property of mustard gas after an acute exposure [27]. However, a study led to new findings suggesting the potential carcinogenic effects in the years after exposure. Using a cohort study of 25 years, we compared the incidence of malignant disorders in 7570 injured patients with mustard gas and 7595 of individuals without a history of exposure.

Also, we defined the risk of cancer incidence for the exposure to mustard gas during the follow-up period. The results showed that the prevalence of cancer significantly increased with the exposure to mustard gas. The ratio of the cancer incidence after exposure was 1.81(1.27–2.56, 95% CI), the age-adjusted incidence ratio was 1.64 (1.15–2.34, 95% CI) [28], and the cancer risk ratio was 2.20 (1.41–2.88, 95% CI). However, this has not thoroughly been proven yet and requires further complementary studies.

OTHER INJURIES OF PULMONARY AIRWAYS AND PARENCHYMA

Mustard gas, based on the amount inhaled, causes damage to the mucosal layers of the airway. The injury starts from the upper airways and, in case of increase in the inhaled dose, may reach to the smaller airways and the terminal ones. The reaction intensity varies from mild to severe. Lung injury can be observed in various forms, including upper and lower airways acute inflammation, discharge from the upper parts of the respiratory system, inflammatory exudate, and formation of pseudomembrane in the tracheobronchial tree.

The injuries intensify slowly over several days. The smokers and those with irritable airways or those with acute viral illness may develop wheezing or asthma after mild injury. Inhalation of large amounts of mustard gas in the acute phase can cause early development of the symptoms and further increase of the symptoms in the chronic phase.

ACCUMULATION OF SECRETIONS IN THE AIRWAYS

As mentioned, one of the effects of mustard gas is creating instability in the surface cells of the airways. Studies on laboratory animals have shown that 6 h after exposure to mustard gas, multiple bleeding spots appear on the bronchial surface, and after 12 h edema and atelectasis develop, which is followed by lymphoid necrosis and decrease in lymphocytes. Therefore, it can be concluded that the mustard gas damages the superficial epithelium of bronchi.

As a result of this incident, the hair cells are not able to remove secretions. Secretions accumulate in the bronchi and initially block the bronchioles' entrance and then block the larger bronchi. The blockage causes lung atelectasis.

Unless very large quantities of chemical are inhaled, most of the injuries by mustard gas are limited to the airways and not to the lung parenchyma. These changes happen mostly in upper airways and to a smaller extent in trachea and smaller bronchi. After the blockade of bronchioles, atelectasis and compensatory hyperinflation occurs. These findings have been confirmed in autopsies of Iranian chemically injured victims. Bronchiectasis and tracheobronchial stenosis are caused in case of severe injury.

Since sulfur mustard is in the form of gas, it can easily reach to the lower respiratory tract with particles size of 1–5 μm. Gough receptors exist in the main airways like trachea and main bronchi, and the contact of mustard gas particles with these receptors causes a train of coughs. The coughs are without sputum at first, but it appears gradually.

Cholinergic property of mustard gas causes excessive secretion of the secretory glands of the airways. The secretion in the acute phase is due to the stimulation of cholinergic receptors of the secretary glands. In a healthy person, airway cilia speed drive the secretions toward the upper airways 600–900 times per minute, but awhile after the inhalation of the gas, coughs become productive (with sputum).

IMPAIRMENT OF VENTILATION AND PERFUSION

Asphyxiation happens in the acute phase of mustard gas inhalation. Lung failure can ultimately lead to death. As mentioned, accumulation of secretions in the airways and the lack of its removal lead to obstruction. Obstruction causes lack of oxygen entrance into the lung alveoli and lack of discharge of carbon dioxide, resulting from the metabolism of the cells.

As a result, a type of hypoventilation syndrome occurs and the $PaCO_2$ increases and PaO_2 decreases. However, findings from animal models have shown that after inhalation of high doses of mustard gas, the breathing rate decreases, which can aggravate hypoventilation resulting from the bronchial obstruction.

AIRWAYS BLEEDING (HEMOPTYSIS)

Mucosal bleeding and bleeding associated with inflammation and epithelial damage happen both in the acute and chronic phases of chemical injury with mustard gas. When bleeding is excessive, combined with lesions of the lungs, it can cause respiratory tract obstruction, which leads to choking.

BIOCHEMICAL ABNORMALITIES OF THE LUNG

Bronchoalveolar lavage of lungs in laboratory animals has shown that following inhalation of mustard gas, gamma glutamyl transpeptidase (GGT) activity, protein concentration, and lactate dehydrogenase (LDH) activity increase. The increase in GGT activity indicates the bronchial epithelial damage, and LDH activity and protein concentration indicate the pulmonary cytotoxic activity. These three are the indications of cytotoxic activity in the lungs and bronchial epithelium damage. Sometimes the damage process halts with the elimination of the cause, and sometimes it does not, leading to the destruction of lung tissue in the long run.

PULMONARY VASCULAR INJURY

Mustard gas can cause apoptosis and necrosis of the endothelial cells of pulmonary artery. Decrease in intracellular ATP level is a sign of apoptosis. In these cells, some bands are formed around the nucleus, which do not exist in the necrotic cells. Complementary studies have shown that N-acetyl cysteine can, to some extent, prevent these destructive effects.

IMPAIRMENT OF PULMONARY IMMUNE CELLS

In comparison to other protected organs, the lungs are mostly in contact with the outside environment. In each minute, 5 L of air enters the lungs, which carries living and nonliving particles into the lungs. However, the trachea and bronchi remain sterile because of the lungs' cell defense mechanisms. There is some evidence suggesting that genetic changes happen in cells exposed to mustard gas and it impairs their function. For this reason, some of the lung complications can result from the dysfunction of immobile or circulating cells of the lung parenchyma.

SIGNS AND SYMPTOMS OF OCULAR LESIONS

Due to the strong lipophilic property of the mustard gas and its high affinity for the lipid layer of the tear, the eyes are the most susceptible organs of the body to this gas. The most common signs and symptoms of eye disease are inflammation of the conjunctiva, local edema including the edema of the eyelids, blepharospasm, and epiphora. Other symptoms like miosis, photophobia, and severe eye pain and headache resulting from it may be observed.

In case of mild contact with mustard, inflammation of the conjunctiva may be delayed up to 48 h, followed by a specific stickiness in follicular margins accompanied by mucus discharge. When the damaged eye is opened for examination, it may have a severe pain and a horizontally congested strip may be seen in the eyeball. These observations, which were reported during WWI, are from the examination of patients in the early stages of injury. A white strip replaces the congested strip in the latter stages, which is an indication of more severe damage. It is reported that ulceration of the cornea is not common in the case of mustard gas contact with the eyes.

However, postmortem examinations in 1986 reported a strip injury in the epithelial layer of cornea of the eye of an Iranian victim, who had been exposed to mustard gas. Acute eye contact damaged the anterior chamber of the eye and lens–iris adhesion occurred. Eye infection is one of the serious consequences following eye contact with mustard gas and can lead to blindness. Epithelial defects lead to an increased susceptibility to bacterial infection (*Pseudomonas aeruginosa*), corneal edema, uveitis, photophobia, and turbidity of the cornea. About 75–90% of people who had contact with mustard gas have some degree of visual impairment. Eye injuries are divided into two major categories of early and late lesions.

Based on intensity, the early lesions are divided into three categories:

1. Mild: This type of lesion is caused by a concentration of 12–70 mg/min/m^3 of mustard gas. The symptoms include mild redness of the eyelids and average congestion of conjunctiva. The cornea is usually involved and patients fully recover after a few days.
2. Moderate: The lesion is caused by a concentration of 100–200 mg/min/m^3 of mustard gas. The symptoms include severe eye pain, dryness, and photophobia, and usually appear after about 6 h. Eyelids inflate and severe blepharospasm is caused, so that the examination of the victim will not be possible without the use of local anesthetic drops. Conjunctival swelling and vasculature hyperemia are observed. In cornea, epithelial edema develops along with point defects, which are mainly in eyelid incision and can be dyed with fluorescein. Epithelial defects sometimes occur because of epithelium separation, and due to the bare stump of the sensory nerves of the cornea, the patient feels a severe pain. Microscopic study of conjunctivitis has shown severe loss of mucin cells and obstruction of its blood vessels resulting from the endothelial cell damage. After 48 h, the severe pain and spasm of the eyelid gradually improve, and the corneal epithelium fully recovers after 4–5 days. Complete remission of symptoms usually occurs within 6 weeks or more.
3. Severe: the injury is observed in those exposed to mustard gas at a higher dose of 200 mg/min/m^3. In this group, in addition to the lesions of the moderate category, involvement of the different layers of the corneal and limbal vessels also happen. In addition to visual lesions, the patients simultaneously suffer from respiratory, digestive, and skin complications. Victims complain from severe eye pain, spasm of the eyelids, and blurred vision.

On examination, the swelling and redness of the eyelids and sometimes blisters are observable. Conjunctival hyperemia and severe swelling can be seen, especially in the gap of the eyelids. The limbal area on the nasotemporal side, necrosis, and conjunctival blood vessel damage appear in the form of white necrotic areas.

Irregularities of the corneal epithelium and stromal edema create an orange peel figure and the sensation of the cornea is destroyed in varying degrees. In this phase, epithelial abnormality develops and the addition of microbes like pseudomonas can cause infectious lesions. In some cases, evisceration of eye may be needed. In this phase, in addition to eyelid, conjunctival, and corneal lesions, miosis and anterior uveitis caused by irritation of the cornea is also seen.

Usually, after 1–2 weeks the swelling of the eyelids, conjunctiva, and cornea subsides and the disease goes into three different directions:

1. Complete recovery: in this group, the acute lesions usually recover within 2–6 weeks, but photophobia (sensitivity to light) remains for a while. Finally, all the symptoms disappear and the patient will not have any problems. It should be noted that at this stage, few serious and early symptoms are observed.
2. Symptoms becoming chronic: In this group, the acute lesions subside, but their side effects remain, including photophobia, dry eye (tear deficiency), foreign body sensation, microscopic scratches on the surface of the cornea, limbal ischemia, and sometimes penetration of the limbal blood vessels to the cornea. Penetration of blood vessels into the cornea cause leakage of exudate in the cornea and create lipoid and amyloid deposits. The surface of cornea becomes irregular and some parts become thin and sometimes extend to the Descemet membrane, and even lead to perforation of the cornea.
   Multiple and recurrent corneal scratches also appear on the surface of cornea. It should be noted that despite the presence of inflammation, because of the necrosis of the limbal blood vessels, the eyes have a normal appearance. In these cases, the appearance of the eyes should not be the base of clinical judgment.
3. Late lesions: This group includes those who recover a few weeks after the initial injury and become asymptomatic. But some years later, sudden redness and photophobia relapses. In this phase, the conjunctiva, cornea, and limbus are involved and is classified as delayed keratitis [29].

Figure 5.1 shows the clinical course and prognosis of ocular lesions after exposure to mustard gas.

A study on the eye effects in a group of 22 individuals, who were soldiers exposed to mustard gas, showed that all of them had different degrees of chronic eye complications and primary conjunctival scars, which often were accompanied with dryness and decrease in visual acuity. Dysplasia was observed in 41% of them [30].

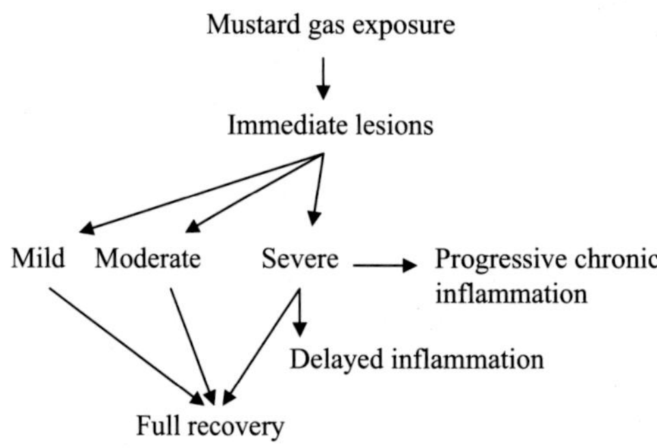

FIGURE 5.1

The clinical course of ocular lesion after exposure to mustard gas.

SIGNS AND SYMPTOMS OF SKIN LESIONS

Skin is one of the first organs that comes in contact with mustard gas, and because of its large contact surface in comparison to other organs, it bears the most damage. Skin complications of mustard gas can be divided into acute and delayed categories. The minimum amount of mustard gas that is necessary to establish the effects of skin erythema is 200 ng/min/m³. This value is affected by factors such as air temperature, humidity, moisture content of skin, and the exposed areas of the body.

Some areas of the body such as the perineum, external genitalia, axilla, the inner part of elbow and neck with thin, warm, and moist skin are sensitive areas.

A drop of liquid sulfur mustard, containing 10 µg of the substance, can cause blisters. About 80% of the amount of sulfur evaporates and only the remaining 10% of it enters the blood circulation. The remaining 1 µg causes blisters. The sensitivity of the skin in individuals with nearly identical skin pigmentation is different. Nevertheless, it seems that people with darker skin are more resistant to mustard gas. Also, frequent exposure to mustard gas increases the sensitivity to it.

Evaporation of small amounts and small droplets happens quickly within 2–3 min, but amounts more than a few hundred milligrams may remain on the skin for several hours. The mustard gas is rapidly absorbed through the skin. The speed at 70°F is 1.4 mg/cm^2/min and at 88°F is 2.7 mg/cm^2/min. After the mustard is absorbed into the skin, it combines with tissue substances and its extraction and separation becomes impossible.

Contact of 50 µg of liquid mustard/cm^2 with skin for 5 min can cause mild erythema; 250–500 µg/cm^2 with the same duration can cause blisters. The lethal dose, for 50% of individuals who are exposed to liquid mustard (LD_{50}), is about 100 mg/kg of the body weight or about 7 mg for a person who weighs 70 kg. This amount equals 1.5 teaspoons of liquid mustard and can cover 25% of the body surface.

Skin contact and chemical reaction with skin proteins weakens cell structure and the extracellular matrix. Also, mustard intensifies skin inflammation by producing the main proinflammatory cytokines such as interleukin-1 beta, interleukin 6, interleukin 8, and tumor necrosis factor alpha. Pigmentation often varies and may increase or decrease. It may develop as a delayed skin disorder in primary sites of mustard lesions. Although the degree of the change of the lesion observed in burn, chemical substances, and other physical damage does not differ, hyperpigmentation is dominant when there is no destruction of the melanocytes.

If melanocytes are destroyed locally and internal migrations of destroyed adnexal structures do not happen, depigmentation prevails. Some of the areas that have already been affected are described as the "sensitive" areas to frequent mechanical damage. These locations may show recurrent blisters after mild injury. In a prospective study, Balali et al. studied the delayed effects of exposure to mustard in a group of Iranian soldiers exposed to mustard gas. The results showed that after 2 years, 41% of the victims had skin pigmentation abnormalities [31].

Occurrence of skin cancers in the location of old scar formation is a known biological phenomenon in the studies of chemical injuries of war. For example, it seems that skin cancers, caused by acute exposure to mustard agent, happen at scar locations; whereas, chronic exposure leads to its formation in any exposed location. Thus, the results of the analysis of the prognosis risk of cancer by the researchers may help, both in chemically injured victims exposed to mustard gas and in the other populations exposed to alkylating agents.

The first and most gentle skin finding is erythema, which appears 2–48 h after exposure and is very similar to sunburn. Symptoms such as itching and burning also appear with it. In skin contact with low amounts of mustard gas, the symptoms can be seen only in the location in contact with the gas. Usually small vesicles appear like a string of pearls around the erythema that may connect to each other and create larger blisters. These blisters appear 1–24 h after exposure of the skin with mustard, although their emergence may be delayed. It seems that after the exposure to liquid mustard, the symptoms appear much quicker than the exposure to mustard gas. In summary, it can be said that erythema occurs 4–8 h after exposure to mustard gas. The blisters appear after 3–18 h, and their completion may take up to several days.

The blisters caused by mustard are particularly thin-walled, dome-shaped, shallow, clear, yellowish, and are surrounded by erythema. Their diameter is between 0.5 and 5 cm, but bigger blisters are probable too. At first, the blister liquid is diluted and clear, or a little opaque. After some time, the color turns to yellow and it becomes more condensed. The blister fluid does not contain mustard and does not act as a stimulant to create blisters. The blisters are not painful by themselves.

They only cause discomfort and may be under pressure. Blisters in the body line surfaces, like the anterior surface of elbows and posterior surface of knees, which may prevent normal mobility. These blisters are delicate and can be easily ruptured by coming in contact with sheets or bandages or during transfer of the victims. New sets of blisters can even appear until the second week after exposure. The burns caused by the exposure to mustard gas are usually first or second grade, but the exposure to liquid mustard causes deeper and even third-degree burns. Exposure to high concentrations of sulfur mustard, like its liquid, usually causes deep burns and leads to the loss of full thickness of the skin. In this case, lesions appear in shape of centers of coagulated necrosis with margins of skin blisters. These lesions are severe and the treatment takes longer, and they are more susceptible to secondary infections. These mostly occur in the genital areas. Necrosis and secondary inflammation are the most common side effect of this type of injury.

The time required for tissue recovery depends on the severity of the injury. Erythema improves after a few days. The recovery of larger lesions, depending on the depth of the lesion, its location, and its extent, may take several weeks to several months; the pace of healing is generally slow.

Parts of the skin that specifically become erythematous are darker, and there may be a sharp increase in pigmentation. Terracotta color to black color may be created in some areas. These changes have been seen specifically in some Iranian casualties and were analyzed and interpreted by Willems. These changes usually disappear within a few weeks and are observed with desquamation of skin layers, leading to the formation of areas of hypopigmentation. The appearance of these areas with parts of the skin with increased pigmentation creates a special figure [32]. Willems corrected the efforts made to connect the extent of the skin lesions with the severity of toxicity. It is very hard to prove this matter. Presence or absence of lung injury and leukopenia is a better guide to determine the degree of toxicity.

Rubbing the damaged skin in patients exposed to mustard can cause secondary blisters. This finding is called the Nikolsky's sign. The sign also occurs in cases of pemphigus vulgaris. The production of secondary blisters is probably a result of rubbing rather than a delayed effect of mustard gas. When the surface skin of the pigmented area is removed, the normal epithelium with normal color appears. It should be noted that blisters are not necessary for the darkness of the skin. Desquamation was reported frequently in the First WWI. However, it was less observed in laboratory studies in which liquid mustard is commonly used.

NONSPECIFIC SYMPTOMS OF MUSTARD GAS

- Distributed skin pigmentation
- Hypotension
- Mental and cognitive disorders
- Loss of sensation and movement of the body
- Speech disorder
- Posttraumatic stress disorder syndrome
- Endocrine gland dysfunction

During WWI, damage to the adrenal glands was used to justify some of these effects. However, this has yet to be verified.

In a case–control study, patients with chemical injury (n = 19) were compared with an age- and sex-matched healthy control group (n = 20). Aerodynamic analysis showed that except for the mean velocity of the flow, a statistically significant difference exists between the two groups in vital capacity, phonation time, phonation volume, the speed of sound, and the total dead volume. This study shows that mustard gas can impair different parameters of aerodynamic speech [33].

In 1986 Norris reassessed some aspects of mustard gas effects and reported the results of research conducted on nervous system function at the end of WWI. In the report, some researchers claimed that they identified the disorder in 22% of patients with mustard gas toxicity [15,34]. He writes, "The researches have reported a type of anxiety in cases of mild toxicity with mustard gas and coughing, photophobia reinforced by hysteria was always observed and infected 22% of the patients with neurosis as follows: functional photophobia in 12.6% of individuals, functional loss of voice in 7.2%, and functional vomiting in 1%." According to a study conducted on chemically injured Iranian, it was found that a high percentage of victims suffer from symptoms of the posttraumatic stress disorder [35].

REFERENCES

[1] Hosseini K, Moradi A, Mansouri A, Vessal K. Pulmonary manifestations of mustard gas injury: a review of 61 cases. Iran J Med Sci 1989;14:20–6.

[2] Papirmeister B, Feister AJ, Robinson SI, Ford RD. Medical defense against mustard gas: toxic mechanisms and pharmacological implications. Boca Raton, FL: CRC Press; 1991.

[3] Keshmiri M. Pulmonary cause of death from chemical warfare agents: the Halabche experience. Iran J Med Sci 1989;14:10–9.

[4] Willems JL. Clinical management of mustard gas casualties. Ann Med Milit Belg 1989;3S:1–61.

[5] Marrs TC, et al. In: Chemical warfare agents: toxicology and treatment, 2nd ed. Chichester, West Sussex, England: John Wiley & Sons, Ltd; 2007.

[6] Drasch G, Kretschmer E, Kauert G, von Meyer L. Concentrations of mustard gas [bis(2-chloroethyl)sulfide] in the tissues of a victim of a vesicant exposure. J Forensic Sci November 1987;32(6):1788–93.

[7] Khateri S, Ghanei M, Keshavarz S, Soroush M, Haines D. Incidence of lung, eye, and skin lesions as late complications in 34,000 Iranians with wartime exposure to mustard agent. J Occup Environ Med November 2003;45(11):1136–43.

[8] Zarchi K, Akbar A, Naieni KH. Long-term pulmonary complications in combatants exposed to mustard gas: a historical cohort study. Int J Epidemiol 2004;33(3):579–81.

[9] Ghanei M, Tazelaar HD, Chilosi M, Harandi AA, Peyman M, Akbari HM, et al. An international collaborative pathologic study of surgical lung biopsies from mustard gas-exposed patients. Respir Med June 2008;102(6):825–30.

[10] Akhavan A, Ajalloueyan M, Ghanei M, Moharamzad Y. Late laryngeal findings in sulfur mustard poisoning. Clin Toxicol (Phila) February 2009;47(2):142–4.

[11] Ghanei M, Akbari Moqadam F, Mohammad MM, Aslani J. Tracheobronchomalacia and air trapping after mustard gas exposure. Am J Respir Crit Care Med February 1, 2006;173(3):304–9.

[12] Ghanei M, Harandi AA, Rezaei F, Vasei A. Sinus CT scan findings in patients with chronic cough following sulfur mustard inhalation: a case-control study. Inhal Toxicol December 2006;18(14):1135–8.

[13] Ghanei M, Naderi M, Kosar AM, Harandi AA, Hopkinson NS, Poursaleh Z. Long-term pulmonary complications of chemical warfare agent exposure in Iraqi Kurdish civilians. Inhal Toxicol August 2010;22(9):719–24.

[14] Ghanei M, Harandi AA. Long term consequences from exposure to sulfur mustard: a review. Inhal Toxicol May 2007;19(5):451–6.

[15] [a] In: Pechura CM, Rall DP, editors. Veterans at risk: the health effects of mustard gas and Lewisite. National Academy Press; 1993.
[b] Ghanei M, Ghayumi M, Ahakzani N, Rezvani O, Jafari M, Ani A, et al. Noninvasive diagnosis of bronchiolitis obliterans due to sulfur mustard exposure: could high-resolution computed tomography give us a clue? Radiol Med April 2010;115(3):413–20.

[16] Ghanei M, Chilosi M, Mohammad HAH, Motiei-Langroudi R, Harandi AA, Shamsaei H, et al. Use of immunohistochemistry techniques in patients exposed to sulphur mustard gas. Pathol Res Int 2011;2011:659–603.

[17] Zarin AA, Behmanesh M, Tavallaei M, Shohrati M, Ghanei M. Overexpression of transforming growth factor (TGF)-betal and TGF-beta3 genes in lung of toxic-inhaled patients. Exp Lung Res June 2010;36(5):284–91.

[18] Mirzamani MS, Nourani MR, Imani Fooladi AA, Zare S, Ebrahimi M, Yazdani S, et al. Increased expression of transforming growth factor-β and receptors in primary human airway fibroblasts from chemical inhalation patients. Iran J Allergy Asthma Immunol May 15, 2013;12(2):144–52.
[19] Ghanei M, Hosseini AR, Arabbaferani Z, Shahkarami E. Evaluation of chronic cough in chemical chronic bronchitis patients. Environ Toxicol Pharmacol July 2005;20(1):6–10.
[20] Aliannejad R, Hashemi-Bajgani SM, Karbasi A, Jafari M, Aslani J, Salehi M, et al. GERD related micro-aspiration in chronic mustard-induced pulmonary disorder. J Res Med Sci August 2012;17(8):777–81.
[21] Karbasi A, Goosheh H, Aliannejad R, Saber H, Salehi M, Jafari M, et al. Pepsin and bile acid concentrations in sputum of mustard gas exposed patients. Saudi J Gastroenterol May–June 2013;19(3):121–5.
[22] Ghanei M, Khedmat H, Mardi F, Hosseini A. Distal esophagitis in patients with mustard-gas induced chronic cough. Dis Esophagus 2006;19(4):285–8.
[23] Shohrati M, Shamspour N, Babaei F, Harandi AA, Mohsenifar A, Aslani J, et al. Evaluation of activity and phenotype of alpha1-antitrypsin in a civil population with respiratory complications following exposure to sulfur mustard 20 years ago. Biomarkers February 2010;15(1):47–51.
[24] Ghanei M, Alikhani S, Mirmohammad SMM, Adibi I, Ramezani T, Aslani J. Evaluation of the presence of emphysema using PFT in comparison with chest HRCT in smokers with a history of exposure to mustard gas. Mil Health J 2007;9(2):139–46.
[25] Sanders C, Nath PH, Bailey WC. Detection of emphysema with computed tomography. Correlation with pulmonary function tests and chest radiography. Invest Radiol April 1988;23(4):262–6.
[26] Uchiyama Y, Katsuragawa S, Abe H, Shiraishi J, Li F, Li Q, et al. Quantitative computerized analysis of diffuse lung disease in high-resolution computed tomography. Med Phys September 2003;30(9):2440–54.
[27] Ghanei M, Harandi AA. Lung carcinogenicity of sulfur mustard. Clin Lung Cancer January 2010;11(1):13–7.
[28] Zafarghandi MR, Soroush MR, Mahmoodi M, Naieni KH, Ardalan A, Dolatyari A, et al. Incidence of cancer in Iranian sulfur mustard exposed veterans: a long-term follow-up cohort study. Cancer Causes Control January 2013;24(1):99–105.
[29] Bagheri MH, Hosseini SK, Mostafavi SH, Alavi SA. High-resolution CT in chronic pulmonary changes after mustard gas exposure. Acta Radiol May 2003;44(3):241–5.
[30] Shohrati M, Peyman M, Peyman A, Davoudi M, Ghanei M. Cutaneous and ocular late complications of sulfur mustard in Iranian veterans. Cutan Ocul Toxicol 2007;26(2):73–81.
[31] Balali-Mood M, Hefazi M, Mahmoudi M, Jalali E, Attaran D, Maleki M, et al. Long-term complications of sulphur mustard poisoning in severely intoxicated Iranian veterans. Fundam Clin Pharmacol December 2005;19(6):713–21.
[32] Ghanei M, Poursaleh Z, Harandi AA, Emadi SE, Emadi SN. Acute and chronic effects of sulfur mustard on the skin: a comprehensive review. Cutan Ocul Toxicol December 2010;29(4):269–77.
[33] Heydari F, Ghanei M. Effects of exposure to sulfur mustard on speech aerodynamics. J Commun Disord May–June 2011;44(3):331–5.
[34] Norris K. Salisbury, UK. Personal communication to author, 1986.
[35] Ahmadi K, Reshadatjoo M, Karami G, Sepehrvand N, Ahmadi P, Bazargan-Hejazi S. Evaluation of secondary post-traumatic stress disorder symptoms in the spouses of chemical warfare victims 20 years after the Iran-Iraq war. Psychiatr Bull 2011;35:168–75.

CHAPTER

Gastroesophageal Reflux in Chemical Patients

Gastroesophageal reflux is a common disorder that causes esophageal and extraesophageal symptoms as a result of stomach content coming up from the stomach into the esophagus. Pulmonary manifestations of reflux are one of most noticeable extraesophageal symptoms [1].

CHRONIC PULMONARY DISEASE AND GASTROESOPHAGEAL REFLUX

The relationship between chronic pulmonary diseases and gastroesophageal reflux has been known for years [2,3]. In fact, gastroesophageal reflux has a high prevalence in different pulmonary diseases, especially in patients with chronic cough, asthma, recurrent pneumonia, chronic obstructive pulmonary disease (COPD), cystic fibrosis, pulmonary idiopathic fibrosis, bronchiolitis obliterans syndrome in lung transplantation, and pharynx and larynx disorders (like laryngitis, hoarseness, pharyngitis, and orodental diseases) [4–6]. In addition, gastroesophageal reflux is a risk factor for the acute exacerbation of COPD and therefore hospitalization and drug administration [7–9]. However, little is known about the pathophysiology of the relationship between pulmonary diseases and gastroesophageal reflux. Moreover, gastroesophageal reflux is associated with the development of bronchiolitis obliterans, pneumonia, and the related diffuse panbronchiolitis in nontransplanted patients [10,11].

MECHANISMS INVOLVED IN THE PATHOGENESIS OF REFLUX AND PULMONARY DISEASES

The suggested mechanisms in the pathogenesis of reflux and pulmonary diseases include microaspiration of the stomach contents into the airways and therefore stimulation and increased resistance of the airways, increased sensitivity and bronchoconstriction mediated by the vagus nerve, and increased sensitivity of the stimulated airways [12–14]. Other mechanisms are inflammatory reactions pathway through IL-18 and recruitment of polymorphonuclear leukocytes as a result of the direct effect of the gastric content on lower airways [15,16]. The bronchial epithelium is known as a source of cytokines, chemokines, and growth factors, and has an important role in the normal pulmonary function. However, different factors like the gastric contents can disturb the integrity of the epithelium. Blondeau et al. reported repeated

Mustard Lung. http://dx.doi.org/10.1016/B978-0-12-803952-6.00006-X

presence of bile acids in the saliva of patients with cystic fibrosis and posttransplantation bronchiolitis obliterans [17].

Recent investigations have revealed that swallowing reflex disorders are frequent in COPD patients and that this finding may be affected by other comorbidities of gastroesophageal reflux [18].

Any disturbance in the complex interaction between swallowing and respiration, and the disorders of swallowing, may increase the risk of aspiration and COPD exacerbation. Studies using submandibular fluoroscopy have shown fluid aspiration during swallowing in COPD patients [19].

In a cohort study, 30 patients with reflux who attended the clinic of Baqiyatallah Hospital, Tehran, Iran, were divided into two groups of with and without the evidence of esophagitis using endoscopy. Spirometry and methacholine challenge tests were performed before and after treatment with proton pump inhibitors. A positive methacholine test was observed in 40% of the patients in the esophagitis group and only 6.7% of the patients in the nonesophagitis group. Treatment with omeprazole was continued for 6 weeks. After the treatment period, marked differences were observed in patients with a positive methacholine test as it decreased from 40% to 13% in the esophagitis group. However, the percentage remained unchanged in the nonesophagitis group [20]. It was interesting that none of the patients had a history of asthma and only the bronchial provocation (challenge) test was positive.

GASTROESOPHAGEAL REFLUX IN CHEMICAL PATIENTS

Reflux has been found as one of the main reasons in the pathogenesis, exacerbation, and continuation of the pulmonary disease in mustard patients. In this regard, a cross-sectional study was performed on mustard patients to evaluate the presence of gastroesophageal reflux through measuring bile acids, pepsin, and trypsin in the bronchoalveolar lavage fluid. The results showed high (more than 8 μmol/L) and low amounts of bile acids in 21% and 53% of the patients, respectively. Only 16% of the patients had no bile acid in the lavage fluid. Trypsin and pepsin were detected in the lavage fluid of all patients [21]. In this study, although all patients received proton pump inhibitors, they suffered from exacerbations of the pulmonary disease with a frequency of more than twice a year or continuous pulmonary symptoms despite receiving the full dose of pulmonary medications. Previous studies have shown that aspiration secondary to gastroesophageal reflux is an important potential factor in the disorders of the transplanted lungs in patients with posttransplantation bronchiolitis obliterans.

A case–control study was conducted to measure and compare the levels of pepsin and bile acids in the sputum of 26 symptomatic mustard-induced bronchiolitis obliterans patients and 12 healthy volunteers. The results showed that the mean pepsin concentration was higher in patients than controls but no difference was observed in bile acids. Table 6.1 shows the results of concentration. Moreover, despite the difference between the groups, the detected differences had no significant correlation with the disease severity [22].

Table 6.1 The Concentration of Pepsin and Bile Acids in Chemical Patients and Healthy Controls

	Control ($n=12$)	Mild MG Exposed ($n=9$)			Moderate MG Exposed ($n=14$)			Severe MG Exposed ($n=3$)		
Factor Analysis	**Mean**	**Mean**	**SD**	**Asymp. Significant**	**Mean**	**SD**	**Asymp. Significant**	**Mean**	**SD**	**Asymp. Significant**
Pepsin in sputum	0.29 ± 0.23	0.215	0.233	0.004	0.287	0.271	0.003	0.312	0.295	0.002
TBA in sputum	0.106 ± 0.19	0.115	0.192	0.731	0.107	0.193	0.901	0.110	0.193	0.623

TBA, total bile acids; *MG*, mustard gas; *SD*, standard deviation; pepsin and TBA unit measurements, respectively, are μg/protein and μg/mg.

In a randomized double-blind placebo-controlled cross-over study, 45 mustard patients suffering from chronic coughs (for 8 weeks or more) and GERD were selected. The patients were divided into two groups. Patients in one group received 20 mg omeprazole twice daily, and patients in the other group received placebo for 4 months. After an interval of one month, the two groups changed their medication for 4 months. The results showed that long-term treatment with high-dose omeprazole, regardless of the reflux symptoms, alleviated chronic coughs and the symptoms of reflux in chemical patients and enhanced the quality of life in both aspects of physical and mental health. However, no marked effects were detected on spirometric indexes in this study. Fig. 6.1 presents the trend of the changes of cough severity in the first 4 months, 1-month drug-free interval, and the second 4 months.

In a study on patients with bronchiolitis obliterans after lung transplantation, increased levels of bile acids in the lavage fluid were reported in 17% of the patients. This finding accompanied the onset of bronchiolitis obliterans in these patients [23]. Moreover, it has been shown that pulmonary aspiration of the gastric content is frequently observed in 50% of these transplanted patients [17]. However, it should be noted that in posttransplantation bronchiolitis obliterans, the vagus nerve and the gastrointestinal tract are manipulated. As a result, the defense system such as the cough reflex and the mucociliary activity undergo changes. In addition, mucociliary clearance decreases very markedly and, therefore, continuous contact with low amounts of the gastric content results in inflammatory reactions and fibrosis [24]. The consumption of immunosuppressive agents like cyclosporine and tacrolimus in transplanted patients also decreases gastric motility and lower esophageal sphincter function [25].

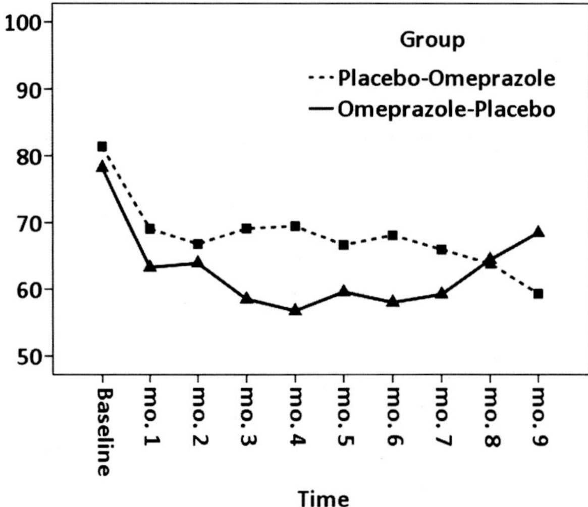

FIGURE 6.1

The trend of the changes of cough severity in the first and second 4 months, indicating improvement in patients receiving omeprazole as compared with the placebo.

Considering the very important role of reflux in pulmonary transplanted patients, Cantu et al., performed early fundoplication surgery, which resulted in marked improvement of reflux and prevention of bronchiolitis obliterans in these patients [26].

A very important point is that the mentioned mechanisms in postpulmonary transplantation bronchiolitis obliterans are not true for isolated bronchiolitis obliterans in mustard patients. COPD patients have many reflux-related symptoms as a result of disorders in the swallowing reflex. These symptoms disturb the inspiration and expiration phase during swallowing. Gastroesophageal reflux and microaspiration are among the causes of bronchiolar and interstitial diseases.

The main limitation of the studies on reflux and aspiration in pulmonary patients is the lack of reliable diagnostic tools because esophageal pH test alone is not enough [27,28].

Because the measurement of impedance measures both acidic and nonacidic episodes of reflux, it might be a superior method for the evaluation of the risk of aspiration; however, this method cannot confirm the presence or lack of aspiration. The best available method for confirmation of esophageal reflux accompanying pulmonary diseases is to treat reflux and evaluate the response to treatment [5].

It has been stated that molecular markers of aspiration like pepsin or bile acids in the saliva or sputum are ideal diagnostic tests. The measurement of pepsin in the saliva or sputum collected during the symptomatic phase of the disease is a sensitive and noninvasive method for detecting gastroesophageal reflux in suspicious patients with atypical signs and symptoms [29]. Although bile acids cause pulmonary injury, their value as a marker of aspiration is a matter of debate [30,31]. A study showed that 42 out of 50 patients with chronic coughs and asthma suffered from gastroesophageal reflux. Of patients with gastroesophageal reflux, 71% of the pulmonary symptoms improved following reflux treatment [32]. Moreover, a randomized clinical trial showed a significant increase in the quality of life and a significant decrease in the severity of pulmonary symptoms following treatment with proton pump inhibitors [33]. However, proton pump inhibitors only decrease the acidity of the stomach while reflux and microaspiration continue to exist. Moreover, proton pump inhibitors increase the risk of community acquired pneumonia and hip fracture [34]. The gold standard for the diagnosis of microaspiration is unknown for the time being; nonetheless, the presence of pepsin and bile acid in the airways is very specific for the diagnosis.

Laboratory studies have shown that chenodeoxycholic acid as part of the bile acids induces the production of TGF-β in the epithelial cells of human airways through p38 MAP-kinase. Moreover, the proliferation of fibroblast cells increases following the exposure to chenodeoxycholic acid. This defensive increase of TGF-β may be a sign of reflux in these patients. TGF-β is a cytokine that plays a role in efferocytosis of the macrophages, inflammation suppression, and regeneration and improvement of the tissues.

A study was conducted on three isoforms of TGF-β using RT-PCR. The results showed that the levels of TGF-β1 and TGF-β3 were significantly higher in chemical patients than the control group while TGF-β2 showed no significant difference [35].

As a result, it has been suggested that TGF-β1 and TGF-β3 may enhance efferocytosis as a defense mechanism and play an important role in the regeneration of the airways in these patients. These properties of TGF-β are probably the reason for the longer lifespan of these patients in comparison with the patients suffering from posttransplantation bronchiolitis obliterans. A study performed on the victims of the September 11, 2001, attacks in the United States confirmed that inhalation of chemical agents produced by fires and destruction of the World Trade Center increased reflux in the survivors [36]. This finding shows that contrary to the previous beliefs that associated reflux with drug side effects, reflux is associated with the inhalation of chemical agents and should receive attention in any chemical event [37]. Moreover, reflux in patients suffering from isolated bronchiolitis obliterans due to exposure to mustard gas is not related to the effective pathogens in lung-transplanted patients and has a different mechanism.

Sulfur mustard may directly damage the gastrointestinal tract via cholinergic effect, inflammatory reaction of the mucosa, and delayed radiometric effects [38]. In addition, mustard gas impairs the protective mechanisms of the lung including the cough reflex and mucociliary clearance to a large extent. It seems that long-term contact of the aspirated material with the lung may result in more injury to the parenchyma. Furthermore, when reflux causes direct pulmonary injury, alloimmune responses may ensue due to the development of an uncontrolled inflammatory environment in these patients [22].

CONCLUSIONS

The exact mechanism of gastroesophageal reflux in mustard patients is not known. Moreover, it is not clear whether gastroesophageal reflux and bronchiolitis obliterans develop simultaneously after the exposure or if one precedes the other and results in the secondary pathology earlier. However, we already know that contact with high-dose mustard gas causes severe injury to the digestive system. Furthermore, it has been shown that the prevalence of both gastroesophageal reflux and esophagitis is higher in patients suffering from bronchiolitis obliterans following exposure to mustard gas than in the normal population [39].

Considering the aforementioned, it can be stated that gastroesophageal reflux is per se an aggravating factor in bronchiolitis obliterans and an important reason for resistance to treatment. Therefore, reflux is no longer considered a comorbidity but a trigger and aggravating factor in the course of the disease.

An important point to remember is that the digestive symptoms of reflux may be very subtle in the patients, leaving us with pulmonary side effects only. This condition is known as silent reflux [40].

In addition to medical therapy, nutritional recommendations should be made according to the diet and eating habits of the patients (see the chapter on treatment for more information). Since the mechanism of reflux is unclear, certain foods and lifestyles with which the patients are satisfied are acceptable, although it may not be possible to discontinue treatment in mustard patients due to recurrence. It is recommended to wait for at least 2 months to see the results of treatment. It is important to provide the patients with adequate information on the disease and its treatment to win their cooperation.

It should be mentioned that treatment with proton pump inhibitors alone cannot control all the complications of reflux. In fact, we cannot eliminate reflux and only try to control it. Moreover, although treatment should minimize the complications, the results of several studies have shown pulmonary complications of reflux as microaspiration despite treatment. More investigations are warranted to understand the accurate pathophysiology and to identify appropriate treatment strategies in this regard.

REFERENCES

[1] Vakil N, van Zanten SV, Kahrilas P, Dent J, Jones R. Global consensus group. The Montreal definition and classification of gastroesophageal reflux disease: a global evidence-based consensus. Am J Gastroenterol 2006;101:1900–20.

[2] Pearson JE, Wilson RS. Diffuse pulmonary fibrosis and hiatus hernia. Thorax 1971;26:300–30.

[3] Belsey R. The pulmonary complications of dysphagia. Thorax 1948;4:44–56.

[4] Feigelson J, Girault F, Pecau Y. Gastro-oesophageal reflux and esophagitis in cystic fibrosis. Acta Paediatr Scand 1987;76:989–90.

[5] Tobin RW, Pope 2nd CE, Pellegrini CA, Emond MJ, Sillery J, Raghu G. Increased prevalence of gastroesophageal reflux in patients with idiopathic pulmonary fibrosis. Am J Respir Crit Care Med 1998;158:1804–8.

[6] Mokhlesi B, Morris AL, Huang CF, Curcio AJ, Barrett TA, Kamp DW. Increased prevalence of gastroesophageal reflux symptoms in patients with COPD. Chest 2001;119:1043–8.

[7] Takada K, Matsumoto S, Hiramatsu T, Kojima E, Iwata S, Shizu M, et al. Relationship between chronic obstructive pulmonary disease and gastroesophageal reflux disease defined by the frequency scale for the symptoms of gastroesophageal reflux disease. Nihon Kokyuki Gakkai Zasshi September 2010;48(9):644–8.

[8] Rogha M, Behravesh B, Pourmoghaddas Z. Association of gastroesophageal reflux disease symptoms with exacerbations of chronic obstructive pulmonary disease. J Gastrointestin Liver Dis 2010;19:253–6.

[9] Rascon-Aguilar IE, Pamer M, Wludyka P, Cury J, Coultas D, Lambiase LR, et al. Role of gastroesophageal reflux symptoms in exacerbations of COPD. Chest October 2006;130(4):1096–101.

[10] Matsuse T, Oka T, Kida K, Fukuchi Y. Importance of diffuse aspiration bronchiolitis caused by chronic occult aspiration in the elderly. Chest 1996;110:1289–93.

[11] Sadoun D, Valeyre D, Cargill J, Volter F, Amouroux J, Battesti JP. Bronchiolitis obliterans with cryptogenetic-like organizing pneumonia. Demonstration of gastro-esophageal reflux in 5 cases. Presse Med 1988;17:2383–5.

[12] Harding SM. Gastroesophageal reflux: a potential asthma trigger. Immunol Allergy Clin North Am 2005;25:131–48.

[13] Harding SM, Guzzo MR, Richter JE. The prevalence of gastroesophageal reflux in asthma patients without reflux symptoms. Am J Respir Crit Care Med 2000;162:34–9.

[14] Schan CA, Harding SM, Haile JM, Bradley LA, Richter JE. Gastroesophageal reflux-induced bronchoconstriction. An intraesophageal acid infusion study using state-of-the-art technology. Chest 1994;106:731–7.

[15] Farrell S, McMaster C, Gibson D, Shields MD, McCallion WA. Pepsin in bronchoalveolar lavage fluid: a specific and sensitive method of diagnosing gastro-oesophageal reflux-related pulmonary aspiration. J Pediatr Surg 2006;41:289–93.

[16] Kiljander TO, Salomaa ER, Hietanen EK, Ovaska J, Helenius H, Liippo K. Gastroesophageal reflux and bronchial responsiveness: correlation and the effect of fundoplication. Respiration 2002;69:434–9.

[17] Blondeau K, Mertens V, Vanaudenaerde BA, Verleden GM, Van Raemdonck DE, Sifrim D, et al. Gastro-oesophageal reflux and gastric aspiration in lung transplant patients with or without chronic rejection. Eur Respir J April 2008;31(4):707–13.

[18] Terada K, Muro S, Ohara T, Kudo M, Ogawa E, Hoshino Y, et al. Abnormal swallowing reflex and COPD exacerbations. Chest February 2010;137(2):326–32.

[19] Cvejic L, Harding R, Churchward T, Turton A, Finlay P, Massey D, et al. Laryngeal penetration and aspiration in individuals with stable COPD. Respirology February 2011;16(2):269–75.

[20] Ghanei M, Poursaleh Z, Harandi AA, Emadi SE, Emadi SN. Acute and chronic effects of sulfur mustard on the skin: a comprehensive review. Cutan Ocul Toxicol December 2010;29(4):269–77.

[21] Aliannejad R, Hashemi-Bajgani SM, Karbasi A, Jafari M, Aslani J, Salehi M, et al. GERD related micro-aspiration in chronic mustard-induced pulmonary disorder. J Res Med Sci August 2012;17(8):777–81.

[22] Karbasi A, Goosheh H, Aliannejad R, Saber H, Salehi M, Jafari M, et al. Pepsin and bile acid concentrations in sputum of mustard gas exposed patients. Saudi J Gastroenterol May–Jun, 2013;19(3):121–5.

[23] D'Ovidio F, Mura M, Tsang M, Waddell TK, Hutcheon MA, Singer LG, et al. Bile acid aspiration and the development of bronchiolitis obliterans after lung transplantation. J Thorac Cardiovasc Surg May 2005;129(5):1144–52.

[24] Hartwig MG, Appel JZ, Davis RD. Antireflux surgery in the setting of lung transplantation: strategies for treating gastroesophageal reflux disease in a high-risk population. Thorac Surg Clin 2005;15:417–27.

[25] Maes BD, Vanwalleghem J, Kuypers D, Ghoos Y, Rutgeerts PJ, Vanrenterghem YF. Differences in gastric motor activity in renal transplant recipients treated with FK-506 versus cyclosporine. Transplantation 1999;68:1482–5.

[26] Cantu 3rd E, Appel 3rd JZ, Hartwig MG, Woreta H, Green C, Messier R, et al. J. Maxwell Chamberlain Memorial Paper. Early fundoplication prevents chronic allograft dysfunction in patients with gastroesophageal reflux disease. Ann Thorac Surg October 2004;78(4):1142–51.

[27] Ravelli AM, Panarotto MB, Verdoni L, Consolati V, Bolognini S. Pulmonary aspiration shown by scintigraphy in gastroesophageal reflux-related respiratory disease. Chest 2006;130:1520–6.

[28] Sheikh S, Allen E, Shell R, Hruschak J, Iram D, Castile R, et al. Chronic aspiration without gastroesophageal reflux as a cause of chronic respiratory symptoms in neurologically normal infants. Chest October 2001;120(4):1190–5.

[29] Kim TH, Lee KJ, Yeo M, Kim DK, Cho SW. Pepsin detection in the sputum/saliva for the diagnosis of gastroesophageal reflux disease in patients with clinically suspected atypical gastroesophageal reflux disease symptoms. Digestion 2008;77:201–6.

[30] D'Ovidio F, Mura M, Ridsdale R, Takahashi H, Waddell TK, Hutcheon M, et al. The effect of reflux and bile acid aspiration on the lung allograft and its surfactant and innate immunity molecules SP-A and SP-D. Am J Transpl August 2006;6(8):1930–8.

[31] Tack J. Review article: the role of bile and pepsin in the pathophysiology and treatment of gastro-oesophageal reflux disease. Aliment Pharmacol Ther 2006;24:10–6.

[32] Schnatz PF, Castell JA, Castell DO. Pulmonary symptoms associated with gastroesophageal reflux: use of ambulatory pH monitoring to diagnose and to direct therapy. Am J Gastroenterol 1996;91:1715–8.

[33] dos Santos LH, Ribeiro IO, Sánchez PG, Hetzel JL, Felicetti JC, Cardoso PF. Evaluation of pantoprazol treatment response of patients with asthma and gastroesophageal reflux: a randomized prospective double-blind placebo-controlled study. J Bras Pneumol 2007;33:119–27.

[34] Lee JS, Collard HR, Raghu G, Sweet MP, Hays SR, Campos GM, et al. Does chronic microaspiration cause idiopathic pulmonary fibrosis? Am J Med April 2010;123(4): 304–11.

[35] Zarin AA, Behmanesh M, Tavallaei M, Shohrati M, Ghanei M. Overexpression of transforming growth factor (TGF)-beta1 and TGF-beta3 genes in lung of toxic-inhaled patients. Exp Lung Res 2010;36:284–91.

[36] Brackbill RM, Thorpe LE, DiGrande L, Perrin M, Sapp 2nd JH, Wu D, et al. Surveillance for World Trade Center disaster health effects among survivors of collapsed and damaged buildings. MMWR Surveill Summ April 7, 2006;55(2):1–18.

[37] Dacre JC, Goldman M. Toxicology and pharmacology of the chemical warfare agent sulfur mustard. Pharmacol Rev 1996;48:289–326.

[38] Ghanei M, Harandi AA, Tazelaar HD. Isolated bronchiolitis obliterans: high incidence and diagnosis following terrorist attacks. Inhal Toxicol April 2012;24(5):340–1.

[39] Ghanei M, Khedmat H, Mardi F, Hosseini A. Distal esophagitis in patients with mustard-gas induced chronic cough. Dis Esophagus 2006;19:285–8.

[40] Galmiche JP, Zerbib F, Varannes S. Review article: respiratory manifestations of gastro-oesophageal reflux disease. Aliment Pharmacol Ther 2008;27:449–64.

CHAPTER

Diagnostic Methods in Chemical Patients

7

It was first believed that the use of chemical weapons was limited to battlefields. However, concerns increased when they were used against civilians in terrorist acts. The Halabja and Sardasht chemical attacks by Iraq on Iran are the most important and disturbing examples in this regard.

The use of chemical weapons introduced new aspects of diagnosis, treatment, and triage in chemical victims. Detection of the exposure and its severity and duration, and follow-up on its adverse effects in symptomatic and asymptomatic individuals, are very important. Many individuals with low-dose exposure and subclinical symptoms have gone undetected who after a while, with nonspecific or advanced symptoms, could not benefit from available treatments. Early diagnosis and treatment is very important to reduce burden of disease in this setting and accordingly different methods of detection and diagnosis have been developed.

LABORATORY FINDINGS

The topic of biomonitoring was put forward that includes the evaluation and analysis of detectable samples in exposed individuals including the blood, urine, fluid aspirated from blisters, and search for their metabolites and hydrolyzed compounds that are produced following bonding with biomolecules.

The metabolites that result from hydrolysis and are mainly excreted in the urine have a short half-life. As a result, there is little time to hand them over to the medical team. However, the metabolites that bond with large biomolecules like proteins and DNA are available for a long time for experimentation. Tissue samples obtained through autopsy or biopsy are also sometimes used. The use of some samples like the urine sample can be carried out at the site of the incident for diagnostic purposes; moreover, it is fast, noninvasive, easy to perform, and does not require specialized education.

These procedures are not only used after the exposure. Preexposure diagnostic procedures and preventive treatments are both used to control the situation and the following courses of action. These actions are executable when the possibility of exposure to chemical weapons is considered or entry to contaminated sites is expected. However, these anticipations are not always possible. Chemical attacks are launched without prior warning to take the enemy by surprise. Most of the available biomedical specimens belong to the chemical victims exposed to sulfur mustard

Mustard Lung. http://dx.doi.org/10.1016/B978-0-12-803952-6.00007-1

during the Iraq–Iran war of 1980–1988. Until 1995, all the experiments on the specimens were limited to the measurement of unmetabolized sulfur mustard or thiodiglycol (TDG). TDG is the result of the hydrolysis of sulfur mustard. The frozen and stored urine and blood samples made it possible to carry out newer analyses with more advanced technologies. These new techniques are both sensitive to track these substances and specific to detect the biomarkers.

The first step in tracing sulfur mustard is to produce the very active ion of sulfonium. Sulfonium is produced following the cyclization of the ethylene group of sulfur mustard. The sulfonium ion readily reacts with nucleophiles, such as water, or can combine with a variety of nucleophilic sites that occur in macromolecules. The resulting compound is capable of producing free metabolites that are measureable in blood, urine, and tissue samples. After absorption, mustard gas undergoes molecular cyclization in the form of ethylene episulfonium ion, and its chloride anion is released. This process is facilitated by heat and by water, a likely explanation for the vulnerability of the warm and moist regions of the body (mucous membranes, eyes, respiratory tract, etc.) to the acute toxic effects of this compound [1]. Cyclization can occur on both sides of the molecule.

According to a report by Boursnell et al., S-labeled sulfur mustard diffuses in the rabbit's body quickly following intravenous injection. Its activity is mostly maintained in the liver, lungs, and kidneys, and about 20% of S is excreted during 12 h. In rodents, the majority of the injected sulfur mustard is excreted in the urine in 72 h [2].

Urine metabolites include TDG (15%), glutamine-bis(b-chloroethylsulfide) conjugates (45%), glutamine-bis(b-chloroethylsulfone) conjugates (7%), bis(b-chloroethylsulfone) (8%), and some cysteine compounds. These findings are compatible with the results of experiments in rodents after intraperitoneal injection [3].

Several metabolites and adducts with proteins and DNA have been proposed as biomarkers of exposure to sulfur mustard. Major urine metabolites of mustard gas are derived from hydrolysis, oxidation, or conjugation with glutathione. The methods of gas chromatography and mass spectrometry are used to detect β-lyase related metabolites in the urine (Table 7.1). The urine metabolites are considered potential specific markers for exposure to sulfur mustard and are detected in high amounts in chemical patients while they are negative in the healthy population.

URINE SAMPLE

As mentioned in Table 7.1, the urine metabolites of sulfur mustard can be measured in the urine sample a few hours after the exposure and up to 10 days. Two metabolites, TDG and thiodiglycol sulfoxide (TGD-sulfoxide), result from hydrolysis, and three other metabolites result from the reaction of sulfur mustard with glutathione [5]. The urine concentration of TDG hydrolysis, a product of mustard gas, is used to confirm a diagnosis of chemical poisoning in hospitalized patients.

The presence of 1,1'-sulfonylbismethylthioethane in the urine, which is a conjugated product of glutathione enzyme, is a more specific marker because it is not found in the samples obtained from nonexposed individuals. Unmetabolized sulfur mustard has been detected in the postmortem tissues and fluids even one week after exposure to sulfur mustard [6].

Table 7.1 Metabolites and Adducts Suggested as Biomarkers of Exposure to Mustard Gas [4]

Possible Biomarker	Structure	Sampling
β-Lyase Metabolites		
1-Methylsulfinyl-2-[2-(methylthio) ethylsulfonyl] ethane (MSMTESE) 1,1'-Sulfonylbis-[2-(methylsulfinyl) ethane] (SBMSE)	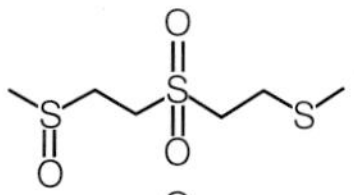 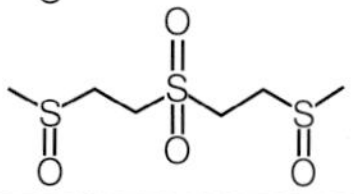	In vivo: detected in human urine 2–11 days after exposure to sulfur mustard (SM)
Hydrolysis and Oxidation Products		
TDG Thiodiglycol sulfoxide (TDGO)	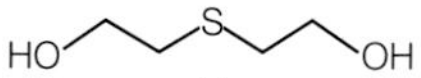 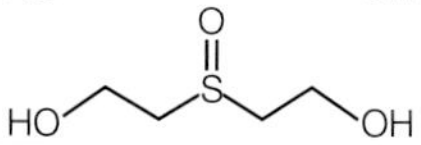 HO S OH HO S(=O) OH	In vivo: detected in human urine 1–11 days after exposure to SM
Hemoglobin Adduct		
N-HETE-valine	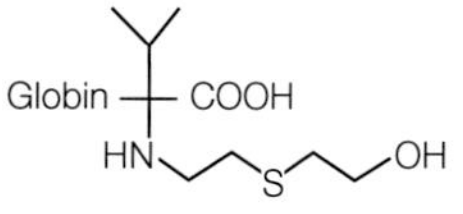 Globin COOH HN S OH	In vivo: detected in rat blood 28 days after the (i.v.) exposure to SM, found also in the blood from human casualties of SM poisoning
Albumin Adduct		
S-HETE-Cys-Pro-Phe	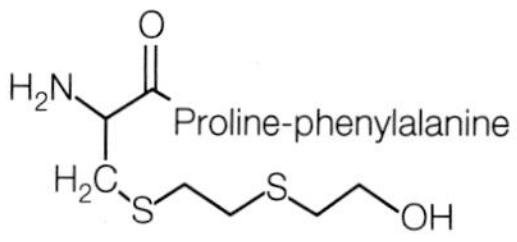 H_2N O Proline-phenylalanine H_2C S S OH	In vivo: detected in human blood 8–9 days after the exposure to SM, and in rat blood 7 days after the exposure
DNA-Adduct		
N7-HETE-Gua		In vivo: detected in urine from guinea pigs 34–48 h after (i.v.) expose to SM
Keratin Adduct		
Nω-HETE-glut or Asp		In vivo: detected in human skin

BLOOD SAMPLES

In general, there are three main methods for investigating the presence of sulfur mustard in blood samples. The first approach includes the evaluation of pure macromolecules like protein and DNA. Sulfur mustard bonded to these molecules can be measured without breaking down macromolecules into smaller fragments. In the second approach, the intended macromolecule is broken into smaller proteins.

Small peptides attached to sulfur mustard are the targets in this method. In the third approach, sulfur mustard attached to macromolecules is completely separated, and sulfur mustard is measured similar to other urine metabolites.

OTHER SAMPLES

The use of other samples like the interstitial tissue, hair, skin, and blister aspirate is not common in comparison with blood and urine samples. These specimens have been mainly used for research purposes. Postmortem evaluations after one year have shown high concentrations of metabolized sulfur mustard in the adipose tissue, skin, brain, and kidneys in a range of 5–15 mg/kg. Moreover, small amounts of sulfur mustard (1–2 mg/kg) have been found in the liver, spleen, and lung tissue. Evaluation of the hair sample one day after exposure has revealed a concentration of 0.5–1.5 mg/kg.

PARACLINICAL FINDINGS

RADIOLOGY

Radiological findings are very helpful in the diagnosis, treatment, follow-up, and evaluation of response to treatment of chemical patients. Chest X-ray and high-resolution computed tomography (HRCT) can be used for radiological evaluation of the lungs.

CHEST X-RAY

Chest X-ray findings of chemical patients are different in symptomatic and asymptomatic conditions. Therefore, detection and interpretation of radiological findings in different situations are crucial in chemical patients. An overview of the radiological findings in symptomatic and asymptomatic patients is presented in the following discussion [7].

Chest X-ray Findings in Asymptomatic Patients

A study showed that patients who were in sulfur mustard–contaminated areas for at least one week and did not exhibit any symptoms of exposure at that time, despite the lack of any clinical symptoms, would develop the complications of exposure. The onset of clinical symptoms varied between 3 years after the presence in the contaminated area to 2 years before the study. None of the patients had abnormal chest X-ray changes, but most of them had abnormal HRCT findings.

Chest X-ray Findings in Symptomatic Patients

Chest X-ray is a conventional diagnostic method that is used in different evaluations. One of the most important applications of chest X-ray is in symptomatic patients in the acute phase after exposure to mustard gas, which is very useful in the evaluation of pneumonia and pulmonary collapse. However, the case is different in chronic conditions. Chest X-ray evaluations of these patients have revealed no mass or nodule and are normal in 70% of the cases. The most common abnormal radiological findings in 34% of the patients are peribronchial infiltration, increased bronchial wall thickness, and increased vascular marking (Table 7.2). Pleural thickening is observed in 5% of the patients. These findings suggest that despite a normal chest X-ray in a large number of patients, many of them suffer from respiratory problems. Therefore, radiography is not a reliable tool to detect and evaluate the lesions of these patients.

In another evaluation, of all 16 patients who had bronchiectatic changes on HRCT, fibrocystic changes of bronchiectasis were observed on chest X-ray images in only three patients. The details of the radiological findings of these patients are presented in Table 7.3.

One study evaluated the chemical patients in the delayed phase and reported that most of them (70%) did not have abnormal radiological findings despite respiratory problems. However, radiography of these patients showed chronic bronchitis changes (15%), reticular pattern (9%), and bronchiectasis (6%).

In another study, radiological findings of mustard patients were hyperinflation (25%), increased bronchovascular thickness (13%), bronchiectasis (9.3%), pneumonic infiltration (7.4%), and pulmonary congestion (3.7%). Most of the bronchiectatic patterns were evident in lower parts of the lungs on radiography. On the other hand, despite abnormal HRCT findings in 98% of the patients, chest X-ray was normal in 56% of them and showed hyperinflation and tracheal abnormalities in the rest of the patients.

From the previous discussion, it can be concluded that the most common radiological finding is a normal chest X-ray and, therefore, a normal chest roentgenogram does not indicate the lack of pulmonary injury. HRCT is recommended if there is any suspicion of the severity of lung involvement [8].

Table 7.2 Chest X-ray Findings in 61 Chemical Patients 2–4 Weeks After Exposure to Mustard Gas

Radiological Findings	Number	Percentage
Normal	31	50
Peribronchiolar infiltration	21	34
Increased interstitial markings	1	3
Transient infiltration (pneumonia)	10	13

Table 7.3 Common Late Chest X-ray Findings of 16 Patients With Bronchiectasis Following Exposure to Mustard Gas

Sign	Number (%)
Increased interstitial markings	13 (81)
Increased bronchiolar wall thickness with or without peribronchiolar infiltration	12 (75)

PULMONARY PERFUSION SCAN

Considering the underlying bronchiolitis and lack of vascular involvement in chemical patients, perfusion scan of the lungs is rarely helpful in these patients. In one study, the pulmonary involvement of 80 mustard patients was evaluated using the perfusion scan, and the results were compared against chest X-ray findings. Fifty-four patients had normal and 26 patients had abnormal perfusion scan results. The most common disorders were multiple segmental defect (20%), nonhomogenous defect (15%), and single nonsegmental defect (8.75%). There were 38 normal and 42 abnormal chest X-ray reports. The most common abnormal radiological finding was hyperaeration. Of 54 patients with abnormal perfusion scan results, chest X-ray was normal in 23 and abnormal in 31 patients. Of 26 patients with normal perfusion scan results, chest X-ray was normal in 15 and abnormal in 11 cases. Moreover, of 38 patients with normal chest X-ray results, perfusion scan was normal in 11 and abnormal in 27 patients. It could be concluded that the results of the pulmonary perfusion scan are normal in most patients and are not diagnostic. When chest X-ray is normal in these patients, perfusion scan may help to detect the pulmonary problem [9].

PULMONARY CT SCAN

The use of CT scan to detect pulmonary lesions is routinely practiced by physicians who work in the field of detection and treatment of the pulmonary complications of mustard gas. CT scan is mostly used to evaluate the pulmonary parenchyma and airway stenosis in two-dimensional and reconstructed sections. An overview of the HRCT findings in symptomatic and asymptomatic patients is presented in the following.

LUNG HRCT IN ASYMPTOMATIC PATIENTS

A study showed that patients who were in sulfur mustard–contaminated areas for at least one week and did not exhibit any symptoms of exposure at that time, despite the lack of any clinical symptoms, would later develop the complications of exposure. The onset of clinical symptoms varied from 3 years after the presence in the contaminated area to 2 years before the study. No positive findings were noted on the HRCT of 38% of the patients, 38% only had air trapping, and the remaining had at least one abnormal HRCT pattern; in other words, 14.7% had bronchial wall thickening, 8.8% had bronchiectasis, and one patient had a mosaic pattern [10]. On CT scan images, areas with air trapping appear darker than the normal tissue. The presence of air trapping of more than 25% is considered the most sensitive and accurate imaging finding for the detection of bronchiolitis obliterans.

HRCT IN SYMPTOMATIC PATIENTS

Bagheri et al. used HRCT to evaluate pulmonary changes following exposure to mustard gas in 50 patients, and its results were compared with clinical and chest X-ray findings. HRCT was abnormal in all patients while chest X-ray was abnormal in only 80% of the patients. Bronchial wall thickening (100%) was the most common positive HRCT finding. Other positive chest X-ray findings were interstitial lung disease (80%), bronchiectasis (26%), and emphysema (24%). No significant relationship was detected between chest X-ray findings and the severity of the pulmonary lesions. Moreover, in patients with normal chest radiography (20%), bronchial wall thickening and interstitial lung disease were reported on HRCT. According to the results of this study, the authors concluded that bronchial wall thickening, interstitial lung disease, and emphysema were the most common radiological findings in these patients. Therefore, they recommended HRCT as the best radiologic diagnostic modality [11].

Hosseini et al. performed bronchography (n=11) and lung HRCT (n=50) to evaluate 61 mustard gas–exposed patients in the delayed phase who mostly (81%) had abnormal spirometric parameters and signs of lung involvement [12]. The anatomical distribution of bronchiectasis in these 16 patients is presented in Table 7.3 and HRCT findings of bronchiectasis are illustrated in Fig. 7.3. Bronchography and HRCT were used to diagnose bronchiectasis in 3 and 13 patients, respectively. Six patients had diffuse and 10 had local pulmonary involvement. The lower lobes were more involved than the upper lobes (10 vs 4). The right middle lobe or the lingual was not involved in any patient. Chest X-ray of most patients did not show bronchiectasis. On HRCT 12 patients (24%) had bronchiectasis that was bilateral in most of them.

Since lung air trapping is observed in 25% of the normal population, the presence of more than 25% air trapping was a diagnostic criterion for bronchiolitis obliterans in this study. The total number of lobes was 259 in all patients. Positive HRCT findings according to the type and number of lesions in all pulmonary lobes were evaluated in all 50 patients; the number of each type of lesion in the involved lobes was determined, and the percentages in all lobes were calculated. Fifty-eight percent of the patients had normal lateral and anteroposterior chest X-ray images, while air trapping and abnormal patterns of larger airways were noted in 44%. In general, 98% of the patients had abnormal HRCT. Significant air trapping (with a frequency of 76% as the most common radiological finding), bronchiectasis, and decreased parenchymal opacity on expiration were the most common abnormal findings (Figs. 7.1 and 7.2). Air trapping alone was the most common finding, detected in 76% of the patients. HRCT of the 10 healthy control subjects was unremarkable except for air trapping in a smoker. Of 250 evaluated lobes, 62% had bronchiectasis, 51% had hyperaeration, 33% had mosaic parenchymal attenuation, and 17% had alveolar septal wall thickening (Table 7.2). Regardless of the type of involvement, the highest and the lowest involvement were observed in the left lower lobe and the

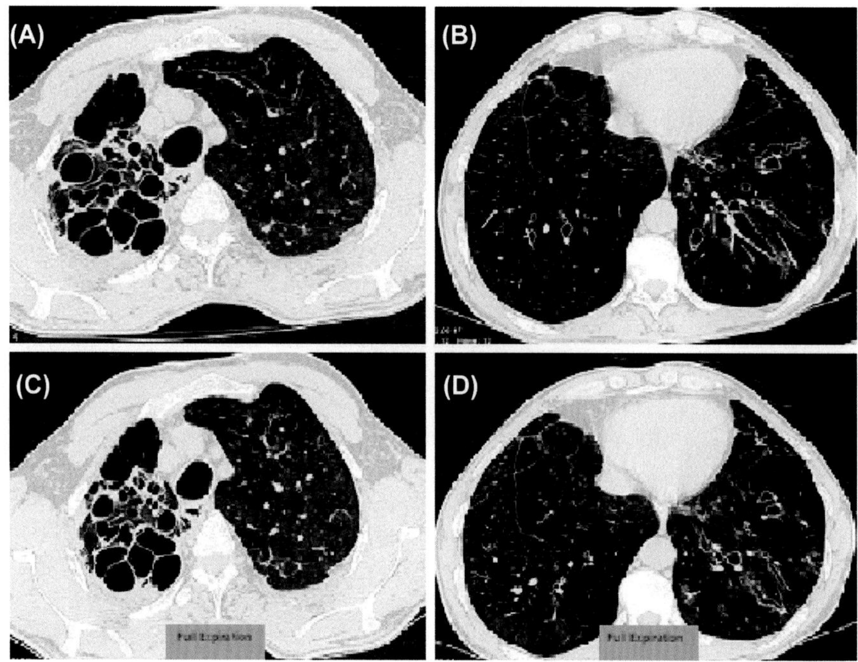

FIGURE 7.1

Cylindrical and cystic patterns of bronchiectasis in the HRCT of some mustard (B and D) and nonmustard patients (A and C).

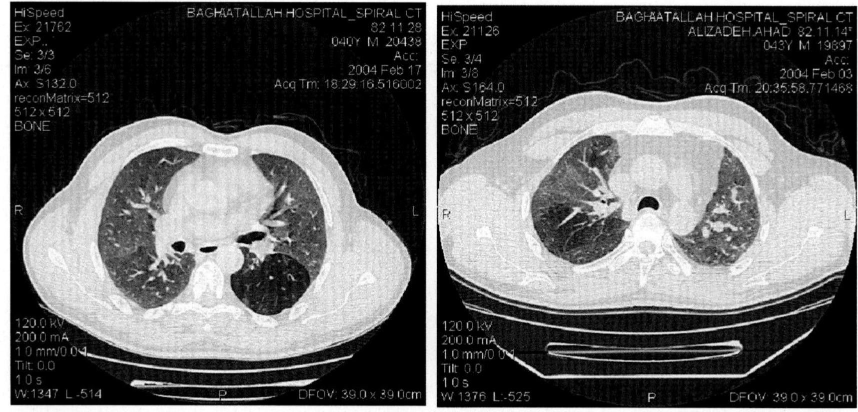

FIGURE 7.2

Air trapping and mosaic parenchymal attenuation of more than 25% in the HRCT of a mustard patient.

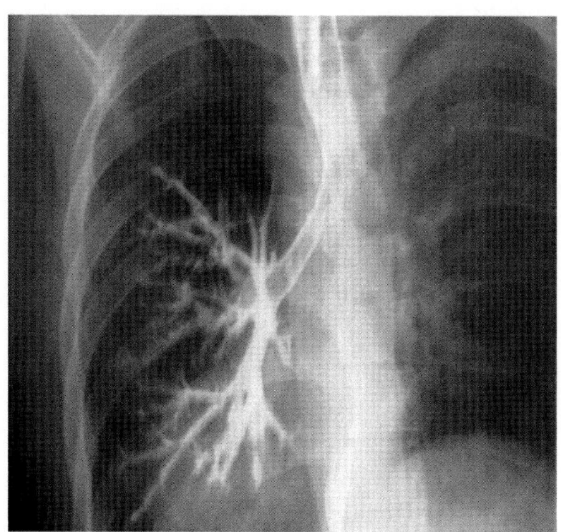

FIGURE 7.3

Left main bronchus stenosis in a bronchogram.

Table 7.4 HRCT Findings in 50 Patients 14 Years After Exposure to Mustard Gas in Comparison With the Control Group

HRCT Findings	Number (%) of the Patients	Number (%) of the Controls	*P* Value
Air trapping	38 (67)	1 (10)	<0.001
Bronchiectasis	37 (74)	0 (0)	<0.001
Mosaic parenchymal attenuation (expiratory)	36 (72)	0 (0)	<0.001
Irregular increase in the diameter of the trachea and major airways	33 (66)	0 (0)	<0.001
Alveolar septal wall thickening	13 (26)	0 (0)	0.1

middle lobe, respectively (Tables 7.4 and 7.5). Moreover, considering different types of pulmonary lesions in different lobes, the frequency of the involvement of different pulmonary lobes was as follows:

Bronchiectasis: right upper and right lower lobes (100%), left upper lobe (95%), left lower lobe (89%), and middle lobe (16%)
Air trapping: left lower lobe (100%), right lower lobe (95%), right upper lobe (61%), left upper lobe (55%), and middle lobe (16%)
Mosaic parenchymal attenuation: left lower lobe (100%), right lower lobe (42%), left upper lobe (39%), and middle lobe (8%)
Secondary lobular abnormalities: left upper lobe (100%), left lower lobe (92%), right upper lobe (77%), and right lower lobe (54%).

Table 7.5 Distribution of the Lesion in Pulmonary Lobes of the Patients 14 Years After Exposure to Mustard Gas Based on the Total Number of Involved Lobes in Each Type of Lesion

HRCT Findings	Involved Lobes						
	Number of Patients/ Involved Lobes	Right Upper Lobe	Middle Lobe	Right Lower Lobe	Left Upper Lobe	Left Lower Lobe	Number (%)
Bronchiectasis	37/185	37	9	37	35	33	151 (62)
Air trapping	38/190	23	6	36	21	38	124 (51)
Mosaic parenchymal attenuation	36/180	13	3	15	14	36	81 (33)
Alveolar septal wall thickening	13/65	13	0	7	10	12	42 (17)
Total lobes		83	18	95	80	119	

In this study, air trapping more than 25% on CT scan was considered a sign of bronchiolitis obliterans, and other causes of bronchiolitis obliterans like the inhalation of toxic gases, smoking, use of penicillamine, collagen vascular diseases, and infections were ruled out. Therefore, mustard gas injury was the only reason for bronchiolitis in these patients.

Moreover, we conducted a study to evaluate whether HRCT can be used as a noninvasive imaging modality to differentiate mustard lung (as a subgroup of bronchiolitis obliterans) from other respiratory diseases. Three groups of patients including mustard patients, patients with severe chronic refractory asthma, and smoker patients were selected. Moreover, 30 nonsmokers were randomly selected as the control group. Pulmonary function testing and HRCT were performed for all participants.

Airway involvement was higher and more frequent than parenchymal involvement in the groups with chemical-induced injury and asthma in comparison with smokers. On the other hand, parenchymal involvement was more frequent than airway involvement in the smokers group in comparison with the other groups. It was concluded that HRCT could be a very useful method for the differentiation of mustard lung, refractory asthma, and pulmonary injury due to long-term smoking [13].

AIRWAY STENOSIS IN PULMONARY SCAN

Different methods have been used to investigate airway stenosis in chemical patients. In a study by Freitag and Firusian in Germany [14], bronchoscopy was introduced as an effective and useful method for the detection and treatment of airway stenosis. Considering the newer methods of evaluation of airway stenosis (like CT scan), the diagnosis of this disorder is now easier and more accurate than before. Moreover, bronchography is one of the useful methods in the diagnosis of airway stenosis.

In a study performed at Isfahan University of Medical Sciences, Isfahan, Iran, CT scan was used to evaluate the status of the trachea of 30 chemical patients who suffered from long-term pulmonary complications. Abnormal tracheal findings were noted in 80 cases (22%). The abnormalities included irregularities of the tracheal wall and tracheal stenosis in the upper parts due to long-term tracheostomy, nodularity of the tracheal wall with a thickness of 2–3 mm, and increased thickness of the cervical trachea (granulation tissue). In this study, it was concluded that despite severe pulmonary problems in these patients, only eight of them had tracheal involvement, which indicates major involvement of the lower airways [15].

Evaluation of 33 mustard patients who were suspicious for stenosis showed airway stenosis in eight of them. The left main bronchus was commonly involved. Fig. 7.3 shows the pattern of one of these patients. This study concluded that CT scan was superior to bronchography in determining the length and severity of the stenosis, especially in the second part of airways [16].

In another study of 300 victims by HRCT, 13 patients had tracheobronchomalacia of whom 11 (85%) had air trapping with a mean score of 5.5 while in the control group, five out of 20 participants (25%) had air trapping with a score of 0.6. Air trapping was significantly more in patients with tracheobronchomalacia. An association was found between the severity of tracheomalacia and air trapping in the case group, but no relationship was detected regarding the severity of bronchomalacia and air trapping. The results showed that air trapping is associated with tracheobronchomalacia, and both of them are observed as long-term complications of mustard injury. Since air trapping strongly suggests chronic bronchiolitis, it can be concluded that chronic bronchiolitis and tracheobronchomalacia develop in a fundamental and effective process in small and large airways, respectively [17].

CARDIOPULMONARY TESTING IN SYMPTOMATIC CHEMICAL PATIENTS

Cardiopulmonary testing is used to evaluate dyspnea in pulmonary patients [18]. This test can be useful when different findings of chemical patients are normal but the patient complains about shortness of breath [19].

There are patients whose spirometry and HRCT are normal. It is very hard to diagnose these patients. For this reason, for the first time we conducted cardiopulmonary testing in patients with exposure to low-dose mustard gas in a case–control study [20]. The case group in this study comprised mustard patients with exertional dyspnea who had unremarkable physical examination, chest X-ray, and spirometry, and their HRCT did not reveal air trapping. Cardiopulmonary testing was performed using the Wasserman protocol. In total, 159 patients in the case group and 10 participants in the control group were evaluated and compared. The results showed that venous oxygen pressure was decreased in chemical patients, but no difference was observed in the cardiopulmonary response to exercise between cases and controls. According to the results of this study, many patients do not have significant laboratory findings despite the involvement of small airways [21]. Although cardiopulmonary testing is recommended in the evaluation of symptomatic patients with normal imaging findings, it is not very useful in differentiating small airway lesions in patients with lung injury due to exposure to low-dose mustard gas.

DIFFERENTIAL DIAGNOSES OF CHRONIC PULMONARY DISEASES RESULTING FROM MUSTARD GAS INJURY

Asthma, reactive airway dysfunction syndrome, emphysema, pulmonary fibrosis, and chronic bronchitis are the most common diseases that should be considered in chemical patients.

ASTHMA AND CHRONIC OBSTRUCTIVE PULMONARY DISEASE

Late pulmonary complications of mustard injury have been evaluated in numerous studies.

In the early years of investigation, asthma, chronic bronchitis, and emphysema were introduced as the most common pulmonary diseases in mustard patients. Imaging and spirometry findings have revealed that chemical patients have degrees of chronic obstructive pulmonary disease, while pathological evaluations do not show emphysema as a dominant finding. Moreover, although reversibility of airway obstruction is a diagnostic criterion of asthma, the reversibility of airway obstruction following the use of bronchodilators is minimal or often absent.

Sandall was one of the first scientists who performed research in this area. He conducted a study on 83 mustard patients in World War I (WWI) and reported the signs of emphysema and chronic bronchitis in 26% and 20% of them, respectively [22]. In the same year, Hankins published another report with similar results [23]. In 1919 Berghoff evaluated the clinical signs of 2000 American soldiers with mustard-induced injury. He reported that 30% and 22% of them had the signs of chronic bronchitis and emphysema about 3–4 months after the injury, respectively [24]. In 1933 Gilchrist et al. assessed the long-term complications of mustard injury in 89 patients after 10 years [25]. The results of the study showed that 27 of them had pulmonary complications like chronic bronchitis and emphysema. Moreover, another study in 1955 on 1267 English veterans of WWI showed that about 80% of the ex-soldiers suffered from chronic bronchitis [26].

A study of chemical patients in 1989 showed that bronchitis and recurrent pneumonia were major pulmonary complications of mustard injury 2 years after the exposure [1]. Ten-year follow-up of a number of Iranian soldiers with mustard injury during the Iraq–Iran war revealed that two-thirds of them had asthma or chronic bronchitis [27].

The World Health Organization (WHO) published a report on the effects of the use of chemical gases by Iraq against Iran in which late complications of mustard

Table 7.6 Late Effects of Mustard Gas on Iranian Victims in a WHO Report (1984)

Injured Organ	Outcome	Percentage
Respiratory system	Chronic bronchitis Asthma Rhinopharyngitis Tracheobronchitis Laryngitis Recurrent pneumonia Bronchiectasis	78%
Central nervous system		25%
Skin		41%
Eye		36.5%

injury were evaluated in a number of victims. According to this report, chronic bronchitis and asthma were observed in 78% of the patients [28] (Table 7.6).

Evaluation of the long-term pulmonary complications of 197 mustard patients after 10 years showed asthma in 10.65%, chronic bronchitis in 57.8%, and fibrosis in 12.18%. A direct correlation was found between the patients' age and the severity of asthma [27]. However, none of the studies observed correct methodology to detect asthma and only used dyspnea, cough, and wheezing to make a diagnosis, while these signs and symptoms are also found in bronchiolitis. The presence of a chronic cough with sputum for 3 months a year for two consecutive years is by definition chronic bronchitis. However, chronic bronchitis cannot explain the obstructive pulmonary lesions and dyspnea. In general, considering the inclusion and exclusion criteria, asthma and emphysema cannot be accepted as the primary consequence of mustard injury.

BRONCHIECTASIS

The WHO reports in 1984 and 1986 mentioned bronchiectasis as one of the long-term complications of mustard injury in Iranian victims. Moreover, based on different studies in mustard patients, bronchiectasis is an important pulmonary consequence of mustard injury; in two studies, 8.6% [27] and 8.8% [10] of the mustard patients had findings related to bronchiectasis. It can be concluded that bronchiectasis is a consequence of chronic bronchiolitis and not a primary consequence of exposure to mustard gas.

PULMONARY FIBROSIS

This disorder is one of the causes of the restrictive pattern in pulmonary function testing of chemical patients. In a study on 197 chemical patients, pulmonary fibrosis was introduced as one of the causes of the restrictive pattern in 12.1% of the patients.

However, pulmonary fibrosis is not a dominant finding in pathological evaluation of the chemical patients [29]. Moreover, complementary studies using the diffusing capacity of the lungs for carbon monoxide and lung HRCT have not confirmed pulmonary fibrosis in these patients [13,30].

BRONCHIOLITIS

After years of research and investigation, inflammation of the bronchioles (bronchiolitis) has been proposed as a persistent pulmonary lesion in mustard patients. In studies by Somani and Babu in 1989, bronchiolitis was reported as a late pulmonary complication in Iranian chemical patients [1]. Moreover, in a report by Willems on the death of an Iranian chemical patient after 185 days, the cause of death after autopsy was found to be severe bronchiolitis. Comprehensive evaluation of the chemical patients shows that bronchiolitis obliterans is the most common pathology and one of the main and progressive pulmonary lesions [29,31].

AIRWAY STENOSIS

Evaluation of the chemical patients two years after mustard injury has shown tracheobronchial stenosis as one of the long-term complications of mustard injury. It should be noted that airway stenosis in chemical patients is different from bronchial stenosis resulting from other causes.

Studies have revealed that contrary to bronchial stenosis caused by long-term intubation or other reasons, the right main bronchus is not involved in this condition, which could be due to the external pressure of the aortic arch on the left main bronchus and its diameter and angle.

In addition, bronchoscopy and CT scan both should be used to detect stenosis to increase the accuracy of diagnosis. The use of bronchoscopy in the diagnosis of subglottic stenosis and stenosis after the second part of the airway is preferred over CT scan, while CT scan can better demonstrate the length and severity of stenosis.

Evaluation of the chemical patients that suffer from airway stenosis and injury shows that their primary pulmonary manifestation is the hemorrhagic inflammation of the tracheobronchial tree, while secondary complications include chronic infection, infectious bronchitis, and severe inflammations that may lead to the stenosis of the trachea and bronchi.

Fifteen years after the primary exposure, scars, wounds, and stenosis progress toward larger airways. In a study evaluating the long-term pulmonary complications of 197 chemical patients 10 years after exposure to mustard gas, 19 patients (9.64%) had airway stenosis with scarring and granulation tissue formation. The stenosis was observed in the trachea, main bronchus, and lobar bronchus in 7, 8, and 4 patients, respectively [17]. Table 7.7 presents the mean age and spirometry results of 17 patients with bronchiectasis and 19 patients with airway stenosis 10 years after exposure to mustard gas.

Table 7.7 Mean Age and Spirometry Results of Chemical Patients 10 Years After Exposure to Mustard Gas

	Age (Year)	FEV_1(L)		FVC(L)		FEV_1/FVC%	PEF rate L/S	
		Observed	%	Observed	%	Observed	Observed	%
Bronchiectasis (n=17)	32.94±1.71	2.22±0.3	65.75	3.01±0.15	70.52	73.59±8.47	5.13±0.75	60.97
Airway stenosis (n=19)	35.94±7.36	1.96±0.26	56.9	2.83±0.26	66.33	67.91±10.56	4.58±0.75	54.41

REACTIVE AIRWAY DYSFUNCTION SYNDROME

Pulmonary hyperreactivity is one of the conditions reported in many studies, even early studies, on long-term complications of mustard injury. Animal studies have shown that reactive airway dysfunction syndrome results from exposure to histamine and substance P. In a study using the in vitro mouse neuroblastoma–rat glioma hybrid NG108-15 clonal cell line model, Ray et al. showed that following exposure to mustard, the amount of the intracellular free calcium ion and arachidonic acid release from the cell membrane increased [32]. These two phenomena could explain the mechanism of reactive airway dysfunction syndrome in mustard patients. In 1933 Matz introduced this consequence in an evaluation of WWI victims after 10 years.

Mustard injury induces long-term sensitivity to cigarette smoke, fumes, air dust, etc., and causes bronchospasm [33]. This pulmonary consequence has received the attention of Iranian researchers, as well.

In a study on 50 nonsmoker chemical patients with a negative history of allergy and overt cardiac symptoms and normal spirograms, methacholine challenge test showed that reactive airway dysfunction syndrome was one of the common pulmonary complications even in patients with normal spirograms [34]. Evaluation of reactive airway dysfunction syndrome in 30 chemical patients who had skin and eye problems with no pulmonary complaints and signs, and comparison of the results with 30 healthy controls (negative history of injury and pulmonary problem), revealed reactive airway dysfunction syndrome in seven (23.3%) and nine (30%) patients who were challenged with cold air and methacholine, respectively, while one participant in the control group had reactive airway dysfunction syndrome. Moreover, it was determined that cold air had a diagnostic value of 93% in detecting reactive airway dysfunction syndrome when compared to methacholine. The results of this study and their comparison with other evaluations in asthmatic chemical patients showed that reactive airway dysfunction syndrome could be one of the long-term complications in any level of pulmonary injury (severe or subclinical) [35].

In evaluation of 17 mustard patients with a diagnosis of asthma, bronchospasm was detected in 82% and 100% of patients after challenge with cold water and methacholine, respectively. Therefore, it is believed that cold water may be a good replacement for methacholine.

In another study, response to cold water and methacholine in 18 mustard patients with chronic bronchitis and 15 participants with no history of pulmonary disease and chemical injury, as the control group, revealed that FEV1 decreased more than 20% in all the patients after methacholine was administered, while 85% of them showed a similar response to cold water. However, the control subjects showed no response to methacholine or cold water.

Recent studies on reactive airway dysfunction syndrome have shown that it is caused by the remodeling of the airways. Airway remodeling is found in chemical patients, as well. These structural changes may result in airway sensitivity clinically. Most of these patients are intolerant to strong odors like the cigarette smoke. The methacholine challenge test may be positive in some of these patients, but it does indicate muscle spasm and can be caused by reasons other than asthma.

EVALUATION OF OTHER PULMONARY COMPLICATIONS

In addition to the already-mentioned diseases, complications like sinusitis, pharyngitis, laryngitis, tracheobronchitis and recurrent bronchitis attacks, tracheomalacia, and, rarely, tracheomegaly have been mentioned as chronic pulmonary disorders associated with mustard injury. These complications could be related directly to mustard inhalation through meticulous evaluation of the articles in this regard, may only be secondary consequences, or may have roots in personal factors (please see the Chapter 9 on the carcinogenesis of mustard gas).

In the end, considering the differential diagnoses and complications of mustard injury, it is suggested that the severity of the lesions and other possible complications should be assessed carefully when evaluating a patient with a history of presence in the contaminated areas. A positive history of exposure to chemical gases, especially any evidence of the development of blisters following the exposure, mandates further evaluations.

DISCUSSION AND CONCLUSIONS

Since chest X-ray is normal in most studies in the acute and chronic phase (up to 80%), its value in the evaluation of lesions in chemical patients as a diagnostic tool is not clear. In fact, because of the weak relationship among the severity of pulmonary lesions, severity of the disorder in pulmonary function tests, and chest X-ray findings, its application in cases other than pneumonia cannot help to detect the lesions in chemical patients. In addition, chest X-ray was mostly used in patients who attended the clinic for numerous complaints; for this reason, it is possible that many patients have normal chest X-ray images. On the other hand, because similar inclusion criteria are not employed in all the studies, it is difficult to make a sound judgment on the percentage of abnormal chest X-ray findings in these patients. In general, bronchial wall thickening, increased interstitial pattern, air trapping and hyperaeration, and disorders of the trachea and bronchiectasis are abnormal findings in the chest X-rays of these patients. In summary, it is better to use other diagnostic modalities like HRCT instead of chest X-ray in these patients to reduce the costs. Considering the pathology of pulmonary diseases in chemical patients, it seems that perfusion scan of the lung is not very useful in these patients while HRCT during full expiration might be very helpful. Since the type and severity of pulmonary lesions are different in the patients in various studies, no definite percentage can be presented for abnormal HRCT findings in these patients. However, bronchial wall thickening, chronic bronchitis, and bronchiectasis are the most common positive HRCT findings in these patients on deep inspiration. Moreover, HRCT results show that the lower lobes are more involved probably due to the increased ventilation of these parts. The conducted studies are not conclusive on the presence of interstitial lung disease. It seems that more studies are required to accept or reject this finding. Recent studies have also revealed that emphysema is uncommon in chemical patients and relate it more to damage to airways as a result of smoking. Researchers believe that the odds of parenchymal injury are low. Since smoking may be one of the causes of emphysema in chemical patients, this point should be considered when interpreting HRCT findings. In general, it seems that pulmonary emphysema solely due to mustard gas is observed in very few patients.

Expiratory HRCT should be performed in patients with probable involvement of small airways because according to the literature, many signs of small airway involvement, like air trapping, are only detectable in the expiratory phase. Mosaic parenchymal attenuation is one of the findings best viewed on expiration. These two signs are the most common abnormal lung HRCT findings of chemical patients in the expiratory phase.

Diseases like bronchiolitis obliterans, asthma, emphysema, bronchiectasis, and chronic bronchitis are associated with air trapping. However, this complication has also been reported in chronic infiltrative lung diseases with small airway involvement (like pulmonary hypersensitivity and sarcoidosis). It should be noted that different levels of air trapping (0–40%) are observed in healthy individuals with no respiratory disease and normal PFT. Lee et al. found air trapping up to 25% in healthy participants [36]. The frequency of air trapping increases with aging, and its severity is associated with smoking. Evaluation of air trapping and estimation of its amount are performed on deep expiration. The patient should hold his breath. Then, HRCT scan is provided in three sections of upper (the upper margin of the aortic arch), middle (apices to main carina), and lower (5–10 cm below the carina) pulmonary regions. Dark areas on the expiratory scan score zero, 1–25% scores 1, 26–50% scores 2, 51–75% scores 3, and 76–100% scores 4.

Therefore, HRCT scans may score 0–24. There are always degrees of air trapping in the upper parts of lower lobes that are normal and should not be regarded in scoring air trapping. Lee et al. found that 5% air trapping is normal in 32% of the healthy population and 5–20% air trapping is normal in 20% of them. In general, a score of 6 or less (25% air trapping or less) can be seen in 52% of the normal population and should not be considered an abnormal condition [36]. Moreover, it has been shown that air trapping more than 32% (a score of 8) following heart and lung transplantation has a specificity and sensitivity of 87.5% for detecting bronchiolitis obliterans. On the other hand, air trapping less than 32% has a high value in ruling out bronchiolitis obliterans in these patients until 5 years after the surgery. In another study, air trapping had a sensitivity of 74% and a specificity of 67% in patients with histopathologically proven bronchiolitis obliterans [37]. These investigations were performed in patients whose disease was confirmed pathologically. In fact, although radiological findings are helpful, they cannot be used to definitely determine small airway disease. On the other hand, since chronic bronchitis, asthma, reactive airway dysfunction syndrome, bronchiectasis, and bronchiolitis obliterans all cause air trapping and are frequently detected in chemical patients, it is not possible to attribute this finding to one of these diseases. To determine the sensitivity and specificity of HRCT findings, they should be compared against the pathological evaluations of the same patients.

Different methods like bronchoscopy, bronchography, and CT scan can be used to detect airway stenosis following the inhalation of high amounts of mustard gas. This finding may be present in different parts of the airway (from the trachea to the bronchi). Based on the conducted studies, the trachea and left main bronchus can be introduced as the most common sites of airway stenosis. Bronchoscopy has a special importance in the diagnosis and treatment of airway stenosis, especially the trachea and main bronchi. However, considering the image reconstruction techniques of CT scan, it can be stated that CT scan is preferred over bronchoscopy in the evaluation of the length and severity of the stenosis, especially in the lower airways.

REFERENCES

[1] Somani SM, Babu SR. Toxicodynamics of sulfur mustard. Int J Clin Pharmacol Ther Toxicol 1989;27:419–35.

[2] Boursnell JC, Cohen JA, Dixon M, Francis GE, Greville GD, Needham DM, et al. Studies on mustard gas (b,b-dichlorodiethylsulphide) and some related compounds. The fate of injected mustard gas (containing radioactive sulphur) in the animal body. Biochem J 1946;40:756–64.

[3] Roberts JJ, Warwick GP. Studies of the mode of action of alkylating agents. The metabolism of bis-2-chloroethylsulphide (mustard gas) and related compounds. Biochem Pharmacol 1963;12:1329–34.

[4] Pesonen M, Vähäkangas K, Halme M, Vanninen P, Seulanto H, Hemmilä M, et al. Capsaicinoids, chloropicrin and sulfur mustard: possibilities for exposure biomarkers. Front Pharmacol December 20, 2010;1:140.

[5] Medical aspects of chemical and biological warfare, Office of the Surgeon General, Department of the Army. [Virtual Naval Hospital, Digital Library of Naval Medicine and Military Medicine; Revision Date: May 1997].

[6] Baselt R. Disposition of toxic drugs and chemicals in man. 9th ed. Seal Beach, CA: Biomedical Publications; 2011. p. 1593–5.
[7] Ghanei M, Harandi AA. Long term consequences from exposure to sulfur mustard: a review. Inhal Toxicol May 2007;19(5):451–6.
[8] Ghanei M, Harandi AA. The respiratory toxicities of mustard gas. IJMS 2010;35:273–80.
[9] Gorji G, Naghiloo A, Elyasi H. Pulmonary perfusion disorders in mustard patients [M.D. thesis]. Shahid Beheshti University of Medical Sciences; 1989–1990.
[10] Ghanei M, Fathi H, Mohammad MM, Aslani J, Nematizadeh F. Long-term respiratory disorders of claimers with subclinical exposure to chemical warfare agents. Inhal Toxicol July 2004;16(8):491–5.
[11] Bagheri MH, Hosseini SK, Mostafavi SH, Alavi SA. High resolution CT in chronic pulmonary changes after mustard gas exposure. Acta Radiol 2003;44(3):241–5.
[12] Hosseini K, Bagheri MH. Development of Bronchiectasis a late sequel of mustard gas exposure. Ir J Med Sci 1998;23:81–4.
[13] Ghanei M, Ghayumi M, Ahakzani N, Rezvani O, Jafari M, Ani A, et al. Noninvasive diagnosis of bronchiolitis obliterans due to sulfur mustard exposure: could high-resolution computed tomography give us a clue? Radiol Med April 2010;115(3):413–20.
[14] Freitag L, Firusian N, Stamatis G, Greschuchna D. The role of bronchoscopy in pulmonary complications due to mustard gas inhalation. Chest November 1991;100(5):1436–41.
[15] Khajavi M. Evaluation of the trachea in chemical patients using CT scan. Isfahan Univ Med J 1997;48:29–35.
[16] Ghanei M, Akhlaghpoor S, Moahammad MM, Aslani J. Tracheobronchial stenosis following sulfur mustard inhalation. Inhal Toxicol December 1, 2004;16(13):845–9.
[17] Ghanei M, Akbari Moqadam F, Mohammad MM, Aslani J. Tracheobronchomalacia and air trapping after mustard gas exposure. Am J Respir Crit Care Med February 1, 2006;173(3):304–9.
[18] Wasserman K, Hansen JE, Sue DY, Whipp BJ, Casaburi R. 3rd ed. Philadelphia: Lippincott Williams and Williams; Principles of exercise testing and interpretation: including pathophysiology and clinical application; p. 199. p.xv.
[19] Chan A, Allen R. Bronchiolitis obliterans: an update. Curr Opin Pulm Med 2004;10:133–41.
[20] Aliannejad R, Saburi A, Ghanei M. Cardiopulmonary exercise test findings in symptomatic mustard gas exposed cases with normal HRCT. Pulm Circ April–June 2013;3(2):414–8.
[21] Beheshti J, Mark EJ, Akbaei H, Aslani J, Ghanei M. Mustard lung secrets: long term clinicopathological study following mustard gas exposure. Pathol Res Pract 2006;202:739–44.
[22] Sandall TE. The later effects of gas poisoning. Lancet 1922;2:857–9.
[23] Hankins JL, Klotz WC. Permanent pulmonary effects of gas in warfare. Am Rev Tuberc 1922;6:571–4.
[24] Berghoff RS. The more common gases: their effect on the respiratory tract. Observation on two thousand cases. Arch Intern Med 1919;24:678–84.
[25] Gilchrist HL, Matz PB. The residual effects of wartime gases. Washington, DC: US Government Printing Office; 1933.
[26] Case RA, Lea AJ. Mustard gas poisoning, chronic bronchitis, and lung cancer; an investigation into the possibility that poisoning by mustard gas in the 1914-18 war might be a factor in the production of neoplasia. Br J Prev Soc Med April 1955;9(2):62–72.

[27] Emad A, Rezaian GR. The diversity of effects of sulphur mustard gas inhalation on respiratory system 10 years after a single heavy exposure: analysis of 197 cases. Chest 1997;112:734–8.
[28] World Health Organization (UN, New York). Report of the mission dispached by the secretary general to investigate allegations of the use of chemical weapons in the conflict between Iran and Iraq. March 12, 1986. Report S/17911.
[29] Ghanei M, Tazelaar HD, Chilosi M, Harandi AA, Peyman M, Akbari HM, et al. An international collaborative pathologic study of surgical lung biopsies from mustard gas-exposed patients. Respir Med June 2008;102(6):825–30.
[30] Ghanei M, Sheyacy M, Abbasi MA, Ani A, Aslani J. Correlation between the degree of air trapping in chest HRCT and cardiopulmonary exercise test parameters: could HRCT be a predictor of disease severity? Arch Iran Med March 2011;14(2):86–90.
[31] Willems JL. Clinical management of mustard gas casualties. Ann Med Milit Belg 1989;3S:1–61.
[32] Ray R, Legere RH, Majerus BJ, Petrali JP. Sulfur mustard-induced increase in intracellular free calcium level and arachidonic acid release from cell membrane. Toxicol Appl Pharmacol March 1995;131(1):44–52.
[33] Veterans at risk: health effects of mustard gas and lewisite. Committee to survey the health effects of mustard gas and lewisite. Washington, DC: National Academy of Sciences, Institute of Medicine, National Academy Press; 1993.
[34] Sohrabpour H, Maleki M. Air way hyperactivity in mustard gas victims with normal spirometry. In: Proceeding of the Sixth Annual Congress on chronic complications of mustard Gas. 1988. Tehran, Iran.
[35] Asad Sanjabi A, Emad A. Evaluation of pulmonary hypersensitivity in mustard patients with ocular and skin problems [M.D. thesis]. Shiraz University of Medical Sciences; 2000.
[36] Lee KW, Chung SY, Yang I, Lee Y, Ko EY, Park MJ. Correlation of aging and smoking with air trapping at thin-section CT of the lung in asymptomatic subjects. Radiology 2000;214:831–6.
[37] Lee ES, Gotway MB, Reddy GP, Golden JA, Keith FM, Webb WR. Early bonchiolitis obliterans following lung transplantation: accuracy of expiratory thin-section CT for diagnosis. Radiology 2000;216:472–7.

CHAPTER

Treatment of Pulmonary Complications in Chemical Patients

8

TREATMENT IN THE ACUTE PHASE

Despite ongoing investigations in the treatment of patients exposed to mustard gas, treatment in the acute phase is still nonspecific and there is no appropriate antidote. The best treatment options in this stage are primary prevention and detoxification [1].

In fact, treatment of the respiratory complications in the acute phase of mustard gas injury is symptom therapy [2]. The main points in the treatment of respiratory problems are the administration of oxygen, addition of humidifiers to ventilation systems, bronchodilators, antibiotics for secondary respiratory infections, and respiratory physiotherapy [3]. Early treatment of respiratory problems resulting from thick sputum includes the administration of mucolytic agents, respiratory physiotherapy, and modification of the body posture. Laryngospasm and stridor may be encountered in acute cases when the patient has been exposed to high doses of sulfur mustard. In these situations, rapid pseudomembrane development in the upper airways may necessitate urgent tracheostomy. Moreover, bronchoscopy may be required to remove the debris and pseudomembrane [4].

Upon the development of chemical pneumonia and acute respiratory distress syndrome, the patient should be hospitalized in the intensive care unit [5]. In severe conditions, fluid and electrolyte correction, and cardiopulmonary support, are among the most important supportive treatments.

In the next steps, scar formation is observed and stenosis may ensue following airway narrowing. In these patients, therapeutic irrigation of the respiratory system with isotonic sodium solution through fiberoptic bronchoscopy is effective and may even decrease mortality. The effects of conventional treatment with antibiotics and corticosteroids to prevent long-term complications are not clear [6]. Animal studies have shown that the use of surfactant, anti-inflammatory agents, and bronchodilators are effective in the treatment of acute pulmonary intoxication with mustard gas [7].

PRIMARY PREVENTION

Exposure to mustard gas results in the development of severe and sometimes lethal complications and causes lifelong disability and numerous problems in different body systems, which impose heavy costs on the treatment and supportive systems of the health care systems in addition to their physical and mental disturbances.

Mustard Lung. http://dx.doi.org/10.1016/B978-0-12-803952-6.00008-3

Therefore, it is evident that primary prevention plays a fundamental role in stopping the development of or decreasing acute and long-term complications.

IDENTIFICATION OF THE CHEMICAL AGENT

The first step in chemical defense is to identify the occurrence of the chemical attack and the employed agent. Common routes of deploying chemical weapons are air bombardment, using missiles and projectiles (shells). Due to their thin shell and the small explosive charge to burst the shell, they produce less sound and destruction power, and fewer pieces of shrapnel than nonchemical bombs. Moreover, following explosion, due to the dispersion of chemical substances, a cloud-like accumulation of toxic gases is formed, which tends to stay close to the earth because it is heavier than the air. Moreover, liquid or solid drops and particles of the chemical agents can be detected on the surface of the earth or equipment. Most chemical gases have certain odors; mustard gas smells like garlic, which distinguishes it from other gases.

Early clinical symptoms are also very important in identification of the type of chemical agent. The presence of respiratory problems and a feeling of choking, redness, and itching of the eyes, and especially skin blisters, are diagnostic keys to mustard gas attack. Detector and alarming kits can also be utilized. Low doses of mustard gas are difficult to detect by humans and do not have the distinctive odor. In these conditions, detector kits are very useful.

There are three protective methods to prevent the effects of chemical agents:

Physical protection includes the use of physical barriers to prevent the penetration of the chemical agent into the body. The protectors include special masks with appropriate filters, special protective clothes, or at least the use of plastic windbreakers, and taking refuge in well-insulated shelters.

Chemical protection includes the use of neutral chemical agents with calcium hypochlorite being the most common. It is a highly oxidizing agent and neutralizes most compounds. Some other neutralizing agents available to the public are perchlorine, lime, or their mixture. It is better to mix these compounds with water.

The best thing to do for mustard decontamination after the use of neutralizing agents is soil excavation to a depth of at least 1–1.5 m in an area of $3 \times 3\ m^2$. Moreover, it is required to decontaminate a perimeter with a radius of at least 10–15 m from the blast zone using neutralizing solutions. The exposed people should be irrigated with copious amounts of water after removing their contaminated clothes. Soap helps with peeling of the superficial skin layer. It is better for the water to contain neutralizing compounds like chloramine-T and sodium thiosulfate [8].

Medical protection includes the treatment of the lesions.

The most important recommendations for primary prevention from the adverse effects of mustard gas are as follows:

- Military training to confront chemical terrorism, especially mustard gas.
- Training for primary preventive measures in the battlefield, especially in chemical wars.

- Training for transferring the victims to the medical care unit with a moist area.
- Using masks with appropriate filters and plastic clothes, preferably with fine texture. Research shows that mustard gas can penetrate leather and ordinary clothes within minutes while rubber and plastic may provide protection for hours [9]. Mustard gas can penetrate the body through ordinary clothes and plastic masks [1].
- Since mustard gas is heavier than the air [10], it is better to transfer the people at risk to an area at least 10 m above the blast area.
- All the contaminated clothes should be removed and disposed of as soon as possible. Since mustard gas can persist in the contaminated clothes and equipment in the fluid form for hours to days and affect biologic tissues, the equipment should undergo detoxification. The equipment that cannot be detoxified should be disposed of in an appropriate fashion [11]. The contaminated material should not be incinerated in the open area to avoid re-release of mustard gas in the air.
- Washing the body with soap as soon as possible is a simple and effective prevention method. It should be mentioned that the soldiers should avoid washing their hands, faces, and bodies with water suspicious of mustard contamination.
- People should be banned from entering the contaminated area because the toxin can persist in the soil for 10 years [1]. Mustard gas can be found at concentrations of 1–25 mg/m^3 in a depth of 6–12 inches in the soil [12].
- Use of standard protection like the mask, clothes, and coverings.
- Sodium hypochlorite and stilbestrol as powder and permanganate can be used for decontamination [11]. If mustard gas is used in the form of powder, decontamination is much more difficult [13].

Mass protection using the shelter allows preservation of the continuous function of people and task forces. The shelter can be temporary or permanent. If people are to stay in the shelter for a long time, an efficient air ventilation system is essential.

Decontamination is different according to the employed chemical weapon. Some chemical compounds like pulmonary agents such as chlorine and phosgene blood gases and nerve gases are unstable and dispersed in the open space and eliminate quickly. However, the conditions are different and very difficult for an agent like sulfur mustard that can persist in the environment for years [8].

TREATMENT IN THE CHRONIC PHASE

After the acute phase, respiratory complications are the most common and most debilitating problems of individuals exposed to mustard gas [14]. According to the published reports, long-term pulmonary complications include airway hypersensitivity, bronchiolitis, bronchitis, bronchiectasis, asthma, tracheal stenosis, cancer, and rarely pulmonary fibrosis [15]. In the past, based on personal experiences, mustard patients were treated like other chronic pulmonary diseases such as asthma and chronic bronchitis, which included symptomatic treatment of chronic complications of mustard exposure. In recent years, guidelines have been published for the treatment of chemical patients according to new evidence derived from complementary studies. In the past, conventional treatment included inhaled corticosteroids and β2 agonists (like beclomethasone and salbutamol), which were prescribed in these conditions for a long time while most of the patients were resistant to nonspecific treatments based on personal experiences. However, special attention is now paid to new findings regarding effective treatments for chemical patients to improve their pulmonary function and restore their normal lives. Complementary studies have shown that the response rate of chemical patients is affected by internal factors (health status, underlying diseases, and genetics) and external factors (occupational exposures before and after exposure to mustard gas, smoking, and exposure to other toxins and pollutants, and number of exposures) [16]. It should be noted that the treatment of chemical patients is in many ways different from the treatment of other pulmonary diseases. It could be argued that the treatment of the pulmonary complications of chemical patients, both in the acute phase of exposure and after that, is the most therapeutic target in these patients.

At first, the physician my believe that they are dealing with a simple known pulmonary disease; therefore, a plan for diagnosis and treatment is designed and followed similar to other pulmonary diseases like asthma and chronic obstructive pulmonary disease (COPD). After a while, due to the adverse effects of the treatments, chronic and debilitating signs and symptoms, and lack of improvement, different aspects of the patient's problems become evident. Since these patients have unique characteristics that cannot be placed in any of the available categories, the word *disorder* was first used for them. However, due to differences between this disease and COPD and asthma, we used the term *mustard lung* to describe chronic pulmonary disease resulting from exposure to mustard gas [17]. This term can, to some extent, explain the specific and unique nature of the pulmonary complications that are caused due to exposure to mustard gas [18].

After it was confirmed that this disease is not similar to other diseases, the results of numerous studies based on radiologic, histologic, and pulmonary function test (PFT) evaluations showed that the most common underlying pathophysiology of this disease was in fact bronchiolitis obliterans, which is different in terms of symptoms and clinical course from bronchiolitis obliterans in children and following pulmonary transplantation. This finding was promising in the beginning because it made it possible to use the available treatments for bronchiolitis obliterans in lung transplantation. However, the differences between these two diseases and therapeutic limitations in transplant patients showed that researchers had a long way ahead for the treatment of mustard lung. The main treatment objectives are to control the symptoms and complaints, to stop the progression of the diseases, and to resolve relative obstruction of the airways. These treatments include a wide spectrum of medical therapy (inhaled or oral), rehabilitation, physiotherapy, and use of oxygen and noninvasive ventilation (NIV). The current treatment strategy is to find the final treatment to regenerate the damaged lung tissue in addition to the herein-mentioned objectives.

We first discuss patients with complaints of chronic cough and dyspnea. In such circumstances, the most important action is a scientific and step-by-step approach to detect the cause of cough and dyspnea.

APPROACH TO PATIENTS WITH COUGH DUE TO CHEMICAL INJURY

After taking a history and physical examination, it is most useful to focus on the anatomic sites of known afferent pathways for cough reflex receptors (the nose, nasopharynx, and lungs).

- In the beginning, chest X-ray should be requested in all patients.
- If history and physical examinations indicate postnasal drip (PND), sinus CT scan is necessary.
- If chest X-ray shows evidence of lung infection, complementary diagnostic methods and sputum study should be requested. If chest X-ray and sinus CT scan are not remarkable, chest high-resolution computed tomography (HRCT) in deep inspiration and expiration should be performed once.
- If past history, physical examinations, and chest X-ray are all normal, the response of the lung function to bronchodilators is evaluated to determine bronchial constriction. All patients should undergo the postbronchodilator test with standard dose of salbutamol or other bronchodilators. Improvement of forced expiratory volume in 1 second (FEV1) by 12% after the administration of the bronchodilator indicates a positive response [19]. It is recommended to try treatment with the maximal dose of the inhaled drug to evaluate the reversibility of the obstructive pulmonary lesion once, maintain the treatment course for at least 2 months and compare the response to treatment before and after drug administration. This finding is very effective in the long-term treatment plan of the patients.
- If the etiology of the cough is not yet determined in this stage, considering the high prevalence of gastroesophageal reflux disease (GERD), standard antireflux treatment for at least 8 weeks is recommended, and the results should be evaluated. It is important not to wait for complaints about reflux like from other patients. Although it has been stated that gastroesophageal endoscopy and pH-metry should be performed to evaluate GERD, empirical treatment and feedback evaluation are more practical. GERD manifests itself with complaints such as heartburn, a sour taste in the mouth, an itchy throat, exacerbation of the symptoms after having fatty or sweet foods, and a feeling of thick and sticky sputum in the morning. Even one or two complaints are enough for diagnostic suspicion. Specific pathologic changes for GERD are observed on endoscopy and of distal esophageal biopsy. Moreover, distal esophageal pH-metry can prove the presence of GERD.
- If the cough persists despite the above-mentioned steps, less common causes like tracheal or bronchial stenosis, bronchiectasis, tumors, etc. should be considered. In this stage, the most appropriate course of action is bronchoscopy to observe the airway above the vocal cords. Meticulous evaluation of the airways for mucosal changes and their shape and also the mucous membrane is essential.

PND is considered in patients with a feeling of mucus accumulating in the throat or dripping from the back of the nose, the need for repeated clearing of the throat, and the presence of mucoid or mucopurulent secretions on physical examination. If there is a repeated need for clearing the throat, if the sinus has a mucosal thickness of more than 6 mm, air-fluid level, or opacity on X-ray or CT scan, rhinosinusitis is suggested as a potential cause of PND. Moreover, it should be kept in mind that if a chemical patient has chronic cough with congestion and purulent secretions on nasal and posterior oropharyngeal examinations with a normal CT scan, GERD should be strongly considered because it can mimic PND.

If patients complain about wheezing, dyspnea, cough, or if wheezing is detected on pulmonary auscultation and spirometric findings indicate the reversibility of airways (more than 12% improvement in FEV1 after salbutamol inhalation), bronchoconstriction should be further evaluated.

The wheezing of bronchiolitis is different from the wheezing of asthma. Although reversibility of wheezing with treatment is possible, a rapid response like in asthma should not be expected.

As mentioned earlier, high-dose treatment should be continued for 2 months. Lack of an appropriate response confirms bronchiolitis obliterans. Maximum-dose treatment is discussed later in the book (recommended prescriptions in the appendix).

After performing necessary diagnostic procedures in patients with chronic cough, different treatment options can be employed according to the cause of chronic cough. A desirable response to a certain treatment indicates the diagnosis of the primary cause and resolving the chronic cough is regarded as an index for response to treatment (complete resolution of cough for two continuous weeks).

It is recommended to request sinus CT scan for the evaluation of the sinuses from the beginning because a normal sinus CT scan rules out a diagnosis of sinusitis and other causes of chronic cough can then be investigated.

Chemical patients may catch pneumonia like other chronic pulmonary patients. So if you suspect pneumonia at any stage, a plain chest X-ray is very helpful. Consider antibiotic therapy if there are abnormal findings in favor of infection.

Other treatments to control respiratory complaints are presented in the following sections. Different drug categories and their indications are discussed later in this chapter and a summary is provided at the end.

MACROLIDES

Alveolar macrophages become in direct contact with mustard gas and are one of the cell lines that are directly affected by mustard gas. In addition to the protective role of macrophages against foreign particles and organisms, they have an important role in the regulation of the humoral immunity through the production of cytokines and activation of T cells. Destruction of the macrophages by mustard gas not only eliminates their phagocytic function but also provides the grounds for pulmonary infections.

Finally, the long-term effects of mustard gas result in the destruction of pulmonary cells and their apoptosis. Macrolides have anti-inflammatory properties through the regulation of inflammatory cytokines. Moreover, complementary studies have shown that they improve the function of macrophages and scavenging of the apoptotic cells. Since the lungs of chemical patients have a noneosinophilic inflammation with the dominance of neutrophils, the role of macrolides in decreasing such an inflammation is more prominent.

Three types of macrolides are used in these patients with published results: azithromycin, clarithromycin, and erythromycin. Studies have shown the effectiveness of these drugs in the acute phase. We also found similar results in our investigations on human subjects in the chronic phase of the disease. It could be stated that macrolides are appropriate treatments in chemical patients from exposure to later stages of the disease.

The effect of macrolides on monocytes exposed to sulfur mustard was investigated in an in vitro study. The results showed that macrolide antibiotics significantly improved monocyte chemotactic and phagocytic activities. Following sulfur mustard (SM) exposure, the chemotactic and phagocytic capacity of the surviving cells was impaired by 41.4% and 26.4%, respectively.

Overexpression of inflammatory cytokines following SM exposure was decreased by 50–70% with macrolide treatment. Moreover, the increased expression of iNOS (iNOS is an important biomarker whose increase results in more production of nitric oxide; it is in fact a bipolar free radical with a short half-life) and overproduction of nitric oxide following cellular exposure to mustard gas was controlled to a large extent. The authors concluded that these changes could finally prevent the damages of the chronic phase and abnormal apoptosis in these cells [20].

Nitric oxide (NO) has a pivotal role in the homeostasis of the nervous, cardiovascular, and immune systems and is effective in the process of wound healing through angiogenesis, cell proliferation, and remodeling of the cellular matrix. Disorders in NO production result in inflammatory processes like arthritis, hepatitis, inflammatory bowel disease, hemolytic shock, and septic shock, while its over-production results in the production of active NO species, like peroxynitrite (sometimes called peroxynitrite), in the presence of O_2 through superoxide anion (O_2^-). This NO metabolite finally causes tissue damage. Moreover, it has been stated that macrolides execute their anti-inflammatory activities through inhibition of nuclear factor-KB (NF-KB).

The proposed therapeutic mechanisms of the effects of macrolides are different for other pulmonary diseases. For example, macrolides activate phagocytosis through the phosphatidylserine receptor in diffuse panbronchiolitis and through collectins in COPD [21,22]. It has been shown that macrolides decrease the volume of sputum and increase its elasticity in patients with diffuse bronchiolitis, chronic bronchitis, and bronchiectasis [23]. Many studies have reported that macrolides reduce the aggregation of neutrophils induced by IL-8, suppress IL-6 expression, decrease superoxide production in neutrophils, and reduce lipopolysaccharide-induced tumor necrosis factor. However, the exact role and mechanism of the effect of macrolides in mustard lung are not yet clear.

The interesting point is that the power and the effect of macrolides seem to be different. For example, Ianaro et al. reported that roxithromycin appeared more effective than erythromycin and clarithromycin, whereas azithromycin only slightly affected the inflammatory reaction in this animal study performed on rats [24].

In one study, N-acetylcysteine and clarithromycin were used in 17 chemical patients for 6 months. The results showed that cough and sputum improved markedly in 85% of the patients. The indexes of FEV_1 and forced vital capacity (FVC) increased by 10.6% and 12.9%, respectively, while FEV_1/FVC did not change significantly [24]. Another study showed similar results for erythromycin. In this study 43 chemical patients received low-dose erythromycin, ie, 400–600 mg daily, for 6 months. The patients were first resistant to bronchodilators and corticosteroids. The 6-month results showed that the variables of dyspnea and the resultant sleep disorders along with hemoptysis decreased significantly in these patients when compared with before the study [25].

ANTIOXIDANTS

A sulfur mustard named 2-chloroethyl ethyl sulfide (CESS) is used in animal studies to cause pulmonary injury to evaluate the pathophysiology of mustard injury and the effect of treatment. (Sulfur mustard, bis-2-chloroethyl sulfide, is a bifunctional alkylating agent while a monofunctional analogue of sulfur mustard, CESS or half mustard, lacks one of the two terminal chlorine molecules.) The interesting point in studies that used SM and CESS was that such vesicants can damage cells by alkylation of macromolecules (ie, DNA, RNA, and proteins), oxidative stress, and glutathione depletion.

Treatment with glutathione, N-acetylcysteine, or tocopherols with glutathione has resulted in the improvement of lung injury in animal studies. Moreover, similar results have been observed following treatment with superoxide dismutase. The basis of all these treatments is oxidative stress involved in mustard injury.

Catalytic metalloporphyrin is an important group of low-molecular-weight antioxidant molecules. One of its types, AEOL 10150, can reduce peroxynitrite and lipid peroxides. Recent in vitro studies have shown that this substance can reduce cytotoxicity and mitochondrial dysfunction 1 h after induced injury by CESS [26,27]. Another study also evaluated the effect of AEOL 10150 in vivo and reported that 18 h after the exposure of laboratory rats to mustard, the level of neutrophils, RBCs, IgM, and the activity of lactate dehydrogenase increased in the bronchoalveolar lavage fluid, but their levels reduced after the administration of AEOL 10150. Moreover, the myeloperoxidase activity was increased in the lung, which was decreased after exposure to AEOL 10150. It also decreased the level of two oxidative stress markers, including plasma 8-hydroxydeoxyguanosine, which were increased following exposure to mustard gas [28].

N-ACETYLCYSTEINE

Sulfur mustard exerts its toxic effects through increasing oxidative stress reactions. One of the involved mechanisms is a decrease in intracellular glutathione (GSH) as an important antioxidant that decreases cellular protection against oxidative stresses. In fact, GSH regulates the level of NO. Stimulation of the macrophages results in the stimulation of nuclear factor–kappa B (NF-KB), which leads to the induction of the production of iNOS and therefore an increased level of intracellular NO.

Based on the theory of oxidative stress, studies were designed to evaluate the effectiveness of antioxidant treatments. In this regard, N-acetylcysteine was used to prevent GSH decrease and control iNOS activity. It should be mentioned that N-acetylcysteine was first used as a mucolytic in pulmonary patients, and its antioxidative and anti-inflammatory properties were noted later. Although in vitro and in vivo studies showed the effectiveness of N-acetylcysteine in the healing of mustard or CESS injuries, its clinical effectiveness had to be confirmed with controlled clinical trials.

In a study conducted by Paromov et al., the laboratory mice injured with CESS received N-acetylcysteine. The results showed a decrease in the indexes of oxidative stress and a final reduction in the toxic effects of mustard [29].

N-acetylcysteine is an antioxidant and a free radical scavenging agent. It is a redox-active agent reported to prevent apoptosis in lymphocytes, neurons, and vascular endothelial cells. In addition, it has a role in the synthesis of glutathione as a thiol-containing compound. Glutathione is a cysteine-containing tripeptide and has an important role in the reduction of free radicals and reactive oxygen. In cells exposed to sulfur mustard, the level of glutathione is decreased and cells become vulnerable to oxidative stress.

The effects of glutathione and sulfur mustard and their relationships have been confirmed in complementary studies.

One piece of evidence in this regard is that the level of glutathione decreased in cells exposed to sulfur mustard; as a result, they became more vulnerable to the cytotoxic effects of sulfur mustard. On the other hand, the cells in which the level of glutathione was increased through N-acetylcysteine were more resistant to the effects of sulfur mustard.

Moreover, it has been shown that N-acetylcysteine, in addition to increasing the level of glutathione, has protective effects against mustard through inhibition of the activation of nuclease factor. Moreover, the effect of N-acetylcysteine was reduced following the synthesis inhibition of glutathione by buthionine sulfoximine, a specific glutathione synthesis inhibitor.

Grandjean et al. showed that the symptoms of bronchitis and chronic pulmonary diseases improved following the use of N-acetylcysteine and their exacerbation was prevented [30]. On the other hand, the lung tissue may be protected with the antioxidant effect of N-acetylcysteine.

Two clinical trial studies evaluated the clinical effectiveness of N-acetylcysteine in chemical patients with normal or impaired PFT. Our study on oxidative enzymes that was conducted years after exposure to mustard gas showed that the problems persisted and oxidants were one of the contributing factors to subsequent disease chronicity.

A study on 250 chemical patients showed that the level of superoxide dismutase and catalase in these patients was higher than in normal people. Moreover, the level of these two factors was higher in patients with more severe injury when compared to patients with milder lesions [31]. Another study showed that the activity of superoxide dismutase, catalase, and glutathione peroxidase in the lavage fluid and erythrocytes was considerably more than the control group. Moreover, the increased activity of glutathione s-transferase in the lavage fluid was associated with decreased glutathione and increased malondialdehyde levels [32]. In addition, the level of glutathione was higher in the control group in comparison with the case group while the level of malondialdehyde was higher in patients [33].

Complementary studies on antioxidant-expressing genes in research centers have confirmed the previously-mentioned findings and have shown that treatment with this group of drugs should be considered in chemical patients.

In a recent study on the glutathione s-transferase gene expression in chemical patients, mRNA was extracted from bronchial specimens provided through bronchoscopy. The results showed overexpression of the glutathione gene to cope with oxidative stress in these patients compared to normal individuals. Moreover, it was found that the expression of heme oxygenase 1, as an antioxidant, was about four times more in chemical patients versus healthy controls [34]. In addition, it has been reported that the superoxide dismutase gene has an increase of about three times while the protein level is decreased [35], which indicates changes in the efficacy of gene translation and posttranslational regulations [36]. Both the protein level and gene expression of MT-1A, which is the most important isoform of metalloprotein, increase upon exposure to free radicals, and they are also reported to be high in chemical patients [37].

According to these findings, we conducted well-designed clinical trials on chemical patients. In the first study, N-acetylcysteine was administered to 72 symptomatic chemical patients with normal PFT at a dose of 1200 mg daily for 4 months, and the results were compared with 72 participants in another group that received placebo [38]. The results showed that dyspnea, wake-up dyspnea, cough, and sputum decreased in patients receiving N-acetylcysteine, and spirometric indexes showed marked improvements. In a similar study with a similar sample size, the effect of a higher dose of N-acetylcysteine (1800 mg/day versus 1200 mg/day) on patients with PFT disorders was investigated. In this study, in addition to improved dyspnea, cough, and sputum volume, PFT indexes showed significant improvements [39]. Studies showed that a combination of clarithromycin and N-acetylcysteine decreased the prevalence of cough and sputum in these patients, and it seems that this drug can effectively reduce the occurrence and exacerbation of bronchiolitis attacks resulting from mustard injury [40]. These findings are in line with previous reports of combination therapy with clarithromycin and N-acetylcysteine in the treatment of chronic bronchitis. In general, we believe that based on the aforementioned studies, this drug should be the basis of treatment and should never be stopped unless there is evidence indicating complete resolution of the problem with other treatment methods.

EVALUATION AND TREATMENT OF GASTROESOPHAGEAL REFLUX

Evaluation of the patients with respiratory problems resulting from exposure to mustard gas indicates a distinct relationship between gastroesophageal reflux and their respiratory problems. The mechanisms involved in the development of these symptoms include bronchial stimulation, microaspiration, and vagally mediated reflex [41]. Our studies showed a marked increase in the prevalence of reflux in these patients [42]. A study on September 11, 2001, patients also showed that inhalation of chemical agents was associated with an increase in reflux [43]. Therefore, contrary to the previous beliefs that associated reflux with drug side effects, it could be related to the inhalation of chemical agents [44]. Our studies revealed that microaspirations following reflux during sleep could explain the progression of bronchiolitis obliterans in chemical patients [45]. From this finding, which is in line with reflux after pulmonary transplantation and its role in the progression of bronchiolitis after transplantation, it can be concluded that gastroesophageal reflux is no longer considered a comorbidity but is an important contributor to the progression of the disease. An important point is that the symptoms of reflux may be very subtle in these patients and only its respiratory complications may be observed, a state known as silent reflux [46]. In these conditions, reflux treatment can still result in marked improvements in respiratory symptoms in patients with chronic pulmonary diseases and empirical treatment with proton pump inhibitors is strongly recommended [46]. The evidence of this disorder has been proved in patients exposed to mustard gas [47]. Gastroesophageal reflux has been reported in 44% of the patients with chronic cough [42].

One of the fundamental questions is the duration of antireflux therapy. Although scientific references recommended 8 weeks of treatment, our daily observations show the recurrence of reflux upon discontinuation of the treatment; therefore, we have had to continue antireflux treatment even for years.

As mentioned earlier, there is a very strong relationship between diet and reflux. The patients should follow an appropriate diet recommended by specialists in order to decrease the severity of the disease and use the minimum dose of the drugs.

Gastroesophageal reflux is a big challenge that requires further research to control.

SHORT-TERM USE OF CORTICOSTEROIDS

It was already known in the late 20th century that the corticosteroids played a role in the improvement of airway obstruction in acute exacerbations of chronic progressive airway diseases. Although some studies like the one conducted by Niewoehner et al. have shown that corticosteroids can improve the signs and symptoms of the diseases and pulmonary function in patients under treatment for 12 h [48], some other studies have reported controversial results [49].

Our study in 2005 showed that short-term intravenous pulse therapy or oral administration of corticosteroids was markedly effective in the exacerbations of chronic bronchitis resulting from mustard gas [50]. However, we found no significant differences between two groups of the study that had different spirometric indexes. According to the results of this study, 13.8% and 30.8% of the patients in each group showed complete or partial response, while no response was observed in the rest of the patients (55.6%) with chronic bronchitis exacerbation. This study showed that it was a chronic obstructive pulmonary disease due to the lack of complete reversibility in pulmonary obstruction. The level of reversibility that was observed in patients was previously reported in patients with chronic obstructive pulmonary patients [51]. The difference in the response of the patients can be related to underlying diseases, previous risk factors, and genetics. However, it should be noted that the inflammatory processes of asthma and bronchiolitis including inflammatory cells, mediators, inflammatory effects, and response to treatment are different. The pattern of inflammation and airway involvement in asthmatic patients is eosinophilic infiltration in all airways without parenchymal involvement. For this reason, it is attributed to hypersensitivity reactions. In chronic pulmonary diseases resulting from mustard gas, lymphocytic and to some extent neutrophilic inflammation is dominant in the airways. In advanced cases, in contrast to asthma, parenchymal destruction is an irreversible damage that results in obstruction during dynamic compression. The eosinophilic inflammation resulting from asthma is apparently suppressed by corticosteroids while these drugs do not show satisfactory results on the inflammatory effects of COPD. According to different studies, long-term oral administration of corticosteroids has failed to stop the progression of the disease while they have confirmed adverse effects. Therefore, they should never be used for a long time in these patients. Any evidence of complete response should raise questions about the primary diagnosis since we may have misdiagnosed a patient with asthma as a case of bronchiolitis. We conducted a pathologic study to differentiate asthma from chronic obstructive bronchiolitis, which confirmed this point [52]. Therefore, the 13.8% complete response rate was due to mild obstruction that approximated the spirometric indexes to the lower limit of normal but could never resolve the complaints completely. Moreover, the 30.8% partial response rate could be explained by COPD in these patients. Considering the fact that the presence of more than 25% air trapping on HRCT in many patients exposed to mustard gas suggests bronchiolitis obliterans, the 50% treatment failure can be due to the coexistence of bronchiolitis obliterans and chronic bronchitis.

From a clinical point of view and based on relative improvement of the patients following treatment with high-dose short-term corticosteroid, the patients can be categorized to responders and nonresponders to steroids. This strategy can help to select the type of treatment. It is recommended that the physician include the results in the patient's record to use them later in case of exacerbation. Since the intravenous pulse of corticosteroids is not more effective than their oral administration, the latter is now preferred to prevent the adverse effects of the intravenous use.

In summary, long-term oral administration of corticosteroids is not recommended in chemical patients. Short-term oral corticosteroid use can be effective in acute exacerbations in half of the patients. Moreover, it should be noted that the corticosteroid use has no effect on the life span of the patients and their various adverse effects cause several problems for the patients. Therefore, it should not be used for more than 1 month and it is recommended to stop its use after 1 week. These recommendations are for oral use of corticosteroids while long-term inhaled steroid use can be effective if indicated.

INHALED CORTICOSTEROID PLUS LONG-ACTING BETA2 AGONISTS

It has been shown that in patients with COPD, a combination of inhaled corticosteroids and long-acting β2 agonists is effective. Since bronchiolitis obliterans belongs to the category of COPD, this combination is more beneficial than inhaled corticosteroid alone. Fluticasone propionate/salmeterol is a combination of inhaled corticosteroid and long-acting β2 agonist that was recently approved in the United States. It is a strong bronchodilator and seems to have important effects on disease exacerbations and the overall quality of life in some patients with COPD.

The objective of treatment in COPD is to prevent and control symptoms and to improve exacerbations. Improvement in the pulmonary function and the health status of the patients are also other objectives of treatment. In patients with moderate to severe COPD, treatment with inhaled fluticasone propionate/salmeterol 50/250 or 50/500 μg twice daily for 24–52 weeks markedly improves pretest FEV_1 when compared to salmeterol alone; it also enhances posttest or FEV_1 after bronchodilator administration when compared to treatment with fluticasone propionate, resulting in a considerable improvement in the quality of life. The use of fluticasone propionate/salmeterol 50/500 μg markedly reduces the incidence of annual COPD exacerbations, especially in severe COPD cases. The dry powder of the inhaled drug as a combination of the corticosteroid and long-acting β2 agonist provides more benefits than monotherapy and can increase the compliance of the COPD patients to use their drugs.

Bronchial abnormalities play an important role in the pathogenesis of chronic pulmonary complications in mustard patients. Bronchiolitis obliterans does not respond to treatment appropriately. The mainstay of its treatment is symptomatic therapy and steroid administration despite debates on the benefits of corticosteroid use. In a study by Wang et al., the use of fluticasone propionate in neonates was not effective in improving acute bronchiolitis [53]. In adults, adding long-acting β2 agonists (salmeterol or formoterol) to inhaled corticosteroids (fluticasone propionate or salmeterol) is more effective than the double dose of inhaled corticosteroid in asthma improvement and COPD control.

We evaluated the effect of fluticasone administration in patients with bronchiolitis obliterans after exposure to mustard gas [54]. The results showed that the use of inhaled corticosteroids plus long-acting β2 agonists was effective in the treatment of bronchiolitis obliterans resulting from exposure to mustard gas. The disease has a reversible nature in about 27% of the patients, and FEV_1 increases in about 10% of the patients during 2 months. It has been observed that hyperresponsiveness of the airways to methacholine in most chemical patients, which is related to FEV_1, may be associated with the chronic inflammation of the airways and hyperresponsiveness of the airways, although about 70% of the chemical patients have shown no response to treatment with inhaled corticosteroids and β agonists. This finding indicates that patients should be categorized according to their response to treatment in order to decrease treatment adverse effects in nonresponders and decrease the costs of treatment.

In phase III of a randomized clinical trial study of 105 patients, the patients received fluticasone propionate/salmeterol 500–1000 μg daily (n=50) or beclomethasone 1000 μg daily plus inhaled salbutamol 800 μg daily (n=53) for 12 weeks. Thirty-six patients in the fluticasone plus salmeterol group and 30 patients in the beclomethasone plus salbutamol group finished the study. After evaluating the PFTs and respiratory symptoms (dyspnea, paroxysmal nocturnal dyspnea, and cough), it was found that both treatment regimens increased FEV1, FVC, FEV1/FVC, and PEF at the end of 12 weeks. The increase in FEV1 was detected in 10% of the patients in the second month.

It was concluded that inhaled corticosteroids and long-acting β2 agonists were effective in the treatment of the patients with bronchiolitis obliterans following exposure to mustard gas. In these patients, in contrast to asthma, the response to treatment should be evaluated after 3 months. Wise drug selection and not imposing the extra costs of new drugs versus older drugs are very important because our results showed that a moderate dose of fluticasone/salmeterol or a combination of high-dose beclomethasone plus a short-acting β agonist had similar effects on airway reversibility [54].

In this study, improvements in FEV1, FVC, FEV1/FVC, and PEF after 3 months of treatment and decreased baseline values during the follow-up period in both groups indicated airway reversibility and effectiveness of spirometry in the follow-up of these patients with similar therapeutic effects for both treatment strategies. It should be mentioned that the dose of administered corticosteroid in the fluticasone/salmeterol group was half of the dose used in the beclomethasone/salbutamol group. Moreover, it should be kept in mind that inhaled corticosteroids have pharmacokinetic differences. The bioavailability of salmeterol and fluticasone is less than beclomethasone dipropionate, resulting in decreased systemic absorption of the swallowed inhaled drug. In higher doses (>1000 μg), budesonide and fluticasone have fewer systemic effects than beclomethasone dipropionate or triamcinolone and are preferred in patients who require high-dose long-term administration of inhaled corticosteroids. It is important that improved dyspnea may compensate for the risks related to long-term corticosteroid use through decreasing the need for steroid use.

As shown in Fig. 8.1, the indexes of PFT had similar changes in both groups, indicating that similar results could be obtained even with a larger sample size. Dyspnea

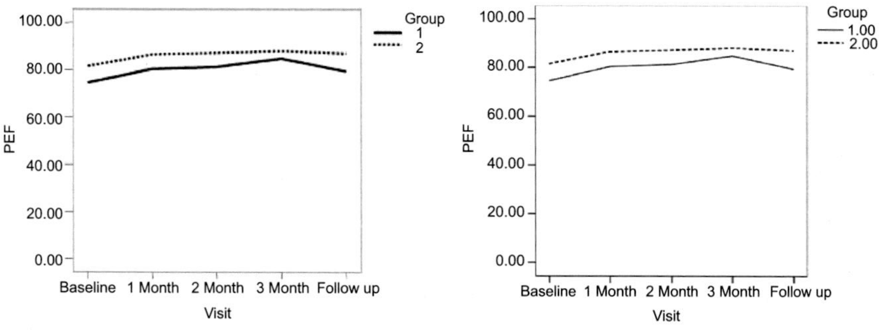

FIGURE 8.1

Comparison of response to treatment in two groups receiving fluticasone/salmeterol and beclomethasone/salbutamol.

improved markedly in the fluticasone/salmeterol group when compared to the beclomethasone/salbutamol group. Moreover, the score of dyspnea returned to baseline after follow-up in both groups. The exacerbation rate was high in the beclomethasone/salbutamol group (27%) while it did not worsen in any of the patients in the fluticasone/salmeterol group. Wake-up dyspnea remained unchanged while dyspnea improved 1 month after the treatment. These findings were contradictory to previous reports indicating improvement in obstructive signs. In this study, the most important causes of the relapse of symptoms were loss to follow-up and the high number of nonresponders (24 patients) in both groups. Although the rate of relapse was lower in the fluticasone/salmeterol group, the difference was not significant.

Considering the emphasis on the benefits of combination therapy in previous studies and also our results, we prefer combination therapy over single therapy in adult patients with bronchiolitis. In general, we recommend a treatment course of 3 months for fluticasone/salmeterol. If patients respond to treatment, they are considered responders and the treatment continues. Other treatment strategies should be administered for other patients (70% in this study); for example, macrolides can reduce both the duration and costs of treatment in these patients. Further studies are warranted to evaluate the long-term effects of treatment with fluticasone/salmeterol on annual attacks and hospitalization rate and total mortality of chemical patients.

THERAPEUTIC EFFECTS OF INHALED ANTICHOLINERGICS

A study by Iranian researchers on the effect of combination therapy with ipratropium bromide and salbutamol versus salbutamol alone in chemical patients showed that combination therapy was more effective than single therapy in the improvement of all parameters including FEV_1, maximal mid-expiratory flow (MMEF), and FVC [55]. When there are concerns about the agonistic effect of the β receptor on the exacerbation of gastroesophageal reflux, inhaled anticholinergic agents can be employed.

In a randomized clinical trial, 54 patients with chronic pulmonary diseases due to exposure to mustard gas were selected and divided into three treatment groups for 2 months (18 patients in each group). The patients in the first group received conventional drugs (Serevent, Flixotide), pulmonary rehabilitation twice a week for 30 min, and tiotropium bromide 18 μg once a day. The patients in the second group received conventional drugs and rehabilitation, and patients in the third group only received conventional drugs.

Cardiopulmonary exercise testing, measurement of pulmonary volumes using plethysmography, and evaluation of the respiratory symptoms were performed before and after the intervention. The results showed that the inhalation of tiotropium bromide as a long-acting bronchodilator in combination with pulmonary rehabilitation (first group) could improve some pulmonary volumes and clinical outcomes in chemical patients. Cough and nocturnal dyspnea improved in the first group versus the third group, and FEV_1, FVC, and MMEF improved in the first group in comparison with the second group. Residual volume increased in the second group, which could be due to dynamic hyperinflation and increased air trapping related to physical activity without using an appropriate bronchodilator. The short-term pulmonary rehabilitation course had no effect on the results of cardiopulmonary exercise testing (the results of the study are not published yet).

OPIOIDS

Since the late 19th century, opioids have been used to alleviate dyspnea in COPD patients through different administration routes. The mechanism of action of nebulized opioid agents is not yet fully understood. They may decrease the feeling of dyspnea through a central effect in the brain. Moreover, it has been shown that even low-dose morphine with a direct effect on the lungs through a nebulizer is effective in some patients. There are three main opioid receptors in the respiratory system including μ (MOR), δ (DOR), and k (KOR). These receptors play a role in the efficacy of three groups of endogenous opioids, ie, endorphins, enkephalins, and dynorphins, and also exogenous opioids like morphine and codeine [56]. In addition, the lung may have new opioid receptors. The k receptor is the dominant receptor in the lung. The suggested mechanism for the therapeutic effects of nebulized morphine is the inhibition of stimulated pulmonary receptors [57].

Opioids depress the release of proinflammatory "substance P" and may help to decrease local inflammation. We also hypothesized that opioids might play a role in decreasing neurogenic inflammation.

In a double blind clinical trial study, we divided 40 patients with a documented history of exposure to sulfur mustard into two groups. The first group received 1 mg morphine sulfate diluted in 4 cc of 0.5% normal saline solution through a nebulizer once daily for 5 days, and the second group, as the control group, received normal saline solution as a placebo. They were visited seven times a day to evaluate the signs, symptoms, and side effects. The scores of dyspnea severity, cough, quality of life, respiratory rate, heart rate, and wake-up dyspnea or cough improved markedly after the use of morphine with no major side effects. Moreover, pick expiratory flow improved considerably following inhaled morphine administration [58].

THERAPEUTIC EFFECTS OF INTERFERON GAMMA

Transforming growth factor beta 1 (TGF-$_{\beta 1}$) is an important pathophysiologic factor in bronchiolitis and peribronchiolar fibrosis. In addition, higher-than-normal levels of TGF-$_{\beta 1}$ have been reported in bronchiolitis and pulmonary fibrosis. On the other hand, INFγ can control this pathway through specific inhibition of the translation of the TGF_{β} gene. Recent studies have shown that patients with moderate clinical complications experience a shift from Th_1 cytokines to Th_2 cytokines if they have decreased levels of INFγ in leukocyte culture.

INFγ and TNF_{β} have opposite roles in collagen synthesis. Once these cytokines are released into the injured tissue from inflammatory cells, their antagonistic reactions play an important role in the regulation of collagen synthesis and maintenance of the homeostasis of the connective tissue.

According to the available clinical data, no reliable and definite immune-compromising treatment regimen exists to improve the pulmonary function in patients with bronchiolitis obliterans. However, the results of the studies support the hypothesis that "the use of $INF_{\gamma/b}$ can improve the pulmonary function in patients with bronchiolitis resulting from exposure to chemical gases" [59]. Moreover, other benefits of the use of $INF_{\gamma/b}$ include shortening the length of hospitalization, increasing arterial O_2, and decreasing dyspnea indexes. Since $TGF_{\beta 1}$ has a pivotal role in the pathogenesis of bronchiolitis obliterans according to very recent studies, we also attribute the improvement in the pulmonary function of the patients to the inhibitory role of INF_{β} on TGF_{β}. TGF_{β} is key member of the cytokines family that controls differentiation and proliferation of different cell lines. Furthermore, studies have shown the ability of TGF_{β} in inducing and transferring the mouse fibroblasts, and we now know that it can affect different cells. In addition, TGF_{β} regulates numerous inflammatory factors that play a role in the development and persistence of pulmonary fibrotic reactions including chemotaxis of macrophages, suppression of the activity of lymphocytes and macrophages, chemotaxis and proliferation of fibroblasts, regulation of collagen synthesis, and the synthesis of the extracellular matrix.

In patients, response to treatment with $INF\gamma_{1b}$ and its effects on the expression of the $TGF_{\beta 1}$ gene can be explained by the downregulation of receptors. Different multicentric studies have shown the effectiveness of $INF\gamma_{1b}$ in the treatment of idiopathic pulmonary fibrosis. However, the effect of $INF\gamma_{1b}$ in chemical patients was evaluated for the first time. Interferon therapy has opened new windows to the pathogenesis of bronchiolitis obliterans. Immunologic advances in the future will find satisfactory solutions to this destructive process. A recent study in our center showed similar results. Following the administration of 100 µg INFγ for 6 months, the quality of life, and the serum levels of IL-2, 4, 6, TNF-α, matrix metallopeptidase-9 (MMP-9), calcitonin gene-related peptide, glutathione, malondialdehyde, catalase, superoxidase, and TGF-β were investigated. The results showed that INFγ improved the quality of life, cough, dyspnea, sputum, and hemoptysis. The serum levels of IL-4, IL-6, IL-10, MMP-9, TNFα, TGF-β, and malondialdehyde were decreased significantly in the end of the study, while the levels of IL-2, IFNγ, and glutathione increased by the end of the trial.

Evidence indicated an improvement in the clinical status of the patients along with desirable and beneficial changes in cytokines and markers of oxidative stress. The drug may not be used routinely because of its very high cost, but it can be effective when the patient has a progressive obstruction that does not respond to other treatments. This drug can be administered in this group of selected patients; consequently, there is an increased chance that the insurance companies and the governments accept to cover the related costs [60].

PROTEASE INHIBITORS

There is evidence of the pathogenic role of proteases like serine, cysteine, and matrix metalloproteinases in injury in animal models exposed to mustard gas [61,62]. Sulfur mustard results in the cellular death of airway macrophages, and the surviving macrophages experience dysfunction, resulting in disturbances in cellular clearance.

Previous studies showed that treatment with aprotinin, ilomastat, and doxycycline reduced the inflammatory response and injury resulting from exposure to mustard gas [63].

Aprotinin forms an irreversible bond with serine-containing proteases and inhibits the function of trypsin, plasmin, kallikrein, elastase, urokinase, and thrombin in a dose-dependent response [64]. Moreover, doxycycline inhibits the nonspecific function of matrix metalloproteinases like MMP-9 and MMP-2 and decreases their cellular and protein levels. In addition, doxycycline and tetracycline result in a decrease in inflammatory reactions and oxidative mediators through affecting the expression of NO [65].

Because we were not able to conduct human studies on these drugs to evaluate their effectiveness in chemical patients due to the side effect of reflux, it is better not to prescribe them in such patients.

OTHER IMMUNE MODULATING DRUGS

Attempts to create reliable and evaluable animal models to investigate contamination with mustard gas have been successful. Basic studies on immune modulators to reverse the suppressing effects of mustard gas have shown interesting results that can be referred to in clinical studies, as well.

The administration of immune-stimulating drugs can effectively reduce severe infections, malignancies, and other complications.

Considering the particular affinity of mustard derivatives for lymphoid cells and DNA molecules, further immunotoxicological studies on sulfur mustard–contaminated patients will shed light on the observed clinical conditions [66].

According to a study by Brockmeyer in 1988, cimetidine, which is extensively used in the treatment of duodenal ulcer in disorders resulting from gastric acid hypersecretory state, can be administered in patients exposed to mustard gas to inhibit the activity of immune suppressing cells and provide the necessary conditions to enhance their immunity [67].

It has been observed that administration of pyrimethamine, which is an antimalaria drug and exerts its effects through inhibition of dihydrofolate reductase enzyme, prepares the grounds for establishment of immunologic capabilities in mustard patients. Immunologic approaches are the subject of future studies for developing effective treatments in these patients.

ALPHA-1 ANTITRYPSIN

Studies have shown that the level of α-1 antitrypsin decreases following exposure to sulfur mustard due to oxidative stress, and there is no specific treatment to prevent the decrease. Therefore, in patients who have coexisting α-1 antitrypsin deficiency and respiratory problems, continuation of the treatment and lifestyle modification can be helpful. Treatments include bronchodilators, inhaled steroid, pneumococcal vaccine, pulmonary rehabilitation, and oxygen administration if required [16].

NONINVASIVE POSITIVE-PRESSURE VENTILATION

Noninvasive positive-pressure ventilation (NIPPV) is a method of respiratory protection via a mask without intubation. In this method, ventilation is provided with a continuous positive pressure or without pressure support in inspiration. A mixture of helium and oxygen is sometimes used in patients with severe but stable pulmonary problems. Heliox, a mixture of helium and oxygen, is a neutral gas, and its density is three times less than the density of air. For its low density, this mixture ventilates the lungs easier with less resistance, resulting in a better laminar flow of air. Studies have shown that the use of this method improves oxygenation and peak expiratory flow, decreases carbon dioxide, and finally reduces respiratory effort and facilitates the delivery of drug aerosols to peripheral alveoli. NIPPV is an effective alternative treatment in severe COPD, bronchiolitis, and other emergency respiratory problems.

One study used this method in chemical patients for the first time [26]. Criteria for initiating NIPPV require at least two of the following: worsening dyspnea during the previous 10 days, respiratory rate >25 breaths/min, arterial pH < 7.35, $PaCO_2 > 50$ mmHg, and $PaO_2 < 50$ mmHg.

Exclusion criteria were recent pneumothorax in the past 1 month, hemodynamic instability or severe respiratory failure requiring intubation, decreased consciousness, lack of patient cooperation, or oral or facial lesions hindering NIPPV.

Twenty-four patients entered the study and received either a mixture of 79% helium and 21% oxygen or a mixture of air and oxygen for 45 min. After that, the patients in each group received the gas mixture of the other group. The results showed that both gas mixtures markedly lowered systolic and diastolic blood pressure, mean arterial pressure, heart rate, respiratory rate, and increased oxygen saturation. The efficacy of heliox was more than the air–oxygen mixture.

TREATMENT OF PULMONARY DISEASE EXACERBATION

Short-term systemic or inhaled corticosteroids, antibiotics, morphine, oxygen, mucolytics, and respiratory physiotherapy can be used in exacerbations. However, the first choice of antibiotic therapy and the duration of treatment are not yet clear [16].

Recent studies have shown that administration of the mixture of helium-oxygen with NIV can decrease airway resistance and improve the work of breathing. The results of the use of heliox have been satisfactory [26].

TERTIARY PREVENTION AND REHABILITATION

When a definite diagnosis is made and the disease is the chronic stage, it is time for tertiary prevention with the objective of stopping the progression of the diseases and reducing disability in the patients. In contrast to morbidity, the mortality rate is low in chemical patients, which results in a large number of patients with advanced and chronic disease in the population. The patients in the advanced stage of the disease benefit from respiratory physiotherapy, respiratory rehabilitation, oxygen, and NIPPV. It should be noted that pulmonary transplantation has no place in these patients because although the patients are severely debilitated, this operation is not necessary for their health. In addition, the results of lung transplantation and its consequences are not yet fully identified.

Exercise intolerance in patients with chronic pulmonary diseases, which mainly results from abnormal ventilation indexes like tidal volume, vital capacity, and maximal voluntary ventilation, considerably affects their health. Any factor that improves these indexes can enhance their efficiency in daily activities. One of these approaches is appropriate physical exercise. However, it should be remembered that exercise cannot resolve structural and physiological abnormalities in patients with obstructive pulmonary disease, although it can increase stamina, strength, efficacy of ventilation, endurance through improving cellular respiration, and adjustment of intracellular structures. Aerobic exercise like lower limb exercises is the mainstay of exercise therapy in chronic pulmonary patients.

Fallah Mohammadi and Tari, in a quasi-experimental study, assessed spirometric indexes in chemical patients with moderate pulmonary disease 3 days before starting an exercise program. Then, to determine the exercise threshold, both groups performed the modified Bruce Protocol Treadmill Stress Test by monitoring heart rate and VO_2max. After that, an exercise program was undertaken for three sessions per week for 4 weeks. In the beginning of each session, the patients did warm-up exercises like walking and stretching and then worked with a bicycle ergometer to gradually increase the intensity of the exercise to 70%. In this study, the effect of 4 weeks of selective aerobic exercise with bicycle ergometer on some spirometric indexes like FEV1, FVC, VC, and TV was evaluated in chemical patients. The results of the study showed that selective aerobic exercise did not have any significant effect on the herein-mentioned indexes, which seems to be due to the short period of the exercise.

In COPD patients, VC is restricted due to the weakness in respiratory muscles, which causes disturbances in full inspiration and leads to expiratory muscle weakness. This disorder also prevents expiration. In addition, the direct effect of decreased muscular force reduces compliance of the lung and chest wall, which results in decreased VC in most patients. It seems that the short period of exercise or differences in type or method of exercise prevented positive effects in enhancing the function of respiratory muscles. One of the limitations of the study was the small sample size of the study. In addition, the participants could not perform long-term exercise due to their special conditions. However, longer exercise programs can resolve ambiguities on the effect of exercise on pulmonary problems. Exercise programs can be expected to be effective if they can reduce mucosal edema to the minimum possible level. Since maximal voluntary ventilation (MVV) and FEV are indexes of the respiratory muscle strength, exercises that strengthen the respiratory muscles improve MVV and FEV, as well. Therefore, specially designed exercise programs can be beneficial in these patients. It seems that performing physical exercise and adapting to it can delay the exacerbation of bronchial inflammation during exercise. Moreover, motivational and psychological conditions could affect the results of spirometry in case and control groups [68].

Considering the new pattern of rehabilitation based on appropriate physical exercise, it is now possible to employ them for optimal results. One exercise limiting factor is the tolerance test that helps to identify the appropriate exercise tolerance range and heart rate. Medical care should be provided at least twice a week, especially in the beginning of the exercise program, to improve the patients' perception and enhance their self-esteem. In addition, the exercise programs should be individualized according to the patients' conditions. Running or intermittent exercise with regular rest intervals gradually make it possible to perform more vigorous exercises. After the first few weeks, the patients may be able to tolerate a high percentage of maximum capacity for 4–24 min per session. Moderate exercises for 17 min at least two days a week are beneficial in most patients because they can reduce the feeling of dyspnea and improve the functional capacity of general well-being of the patients. The exercise should be preferably accompanied by the use of bronchodilators to increase exercise tolerance capacity. The physical exercise should be performed to strengthen all big muscles and comprise aerobic exercise that engages big muscles of the lower limbs. For this purpose, daily activities like walking, hiking, or climbing up the stairs are generally advised for 4–24 min. Since many daily activities require the use of upper limbs, endurance and strength exercises of the upper limbs can be useful, as well [69].

THE EFFECT OF NUTRITION ON DECREASING PULMONARY COMPLICATIONS

Naturally, the lungs are loaded with oxygen and there is a balance between the toxicity of the oxidants (production through the normal cellular function or exposure to peroxidases) and protective activities of several intracellular and extracellular antioxidant defense systems. Imbalance in this system through increased oxidative stress or decreased use of antioxidant resources can trigger a series of pathophysiologic events in the lungs resulting in cell death and pulmonary dysfunction. Cigarette smoke and air pollutants like ozone and nitrogen oxide are rich in free radicals and exposure to them can result in pulmonary oxidative injury through direct oxidation of the pulmonary tissue. Moreover, the process continues through endogenous oxidants, proteolytic enzymes, and activation of neutrophils. Exposure to mustard gas is a typical and extreme example of direct and indirect injury. Furthermore, it is believed that a diet poor in antioxidants like omega 3, carotene, and vitamins C and E may weaken the natural defense and increase the vulnerability to oxidative injury, inflammation, and airway injury. The effect of nutrition on obstructive lung diseases, especially vitamins C and E, has been proved in different studies. Data also show that omega 3 fatty acids may have a potentially protective role against airways hyperresponsivity and decreased pulmonary function. Although epidemiologic data show that consumption of fresh fruits decreases the risk of noncarcinogenic airway limitation, the most appropriate and effective nutrient is not yet identified. Considering nutritional correlations among antioxidant vitamins, especially vitamin C, beta-carotene, flavonoids, and other micronutrients, it is difficult to determine their individual effects.

Some population subgroups with higher levels of oxidative stress like smokers or chemical patients may benefit more from nutritional supplements. Consumption of fresh fruits that are rich in vitamin C has positive effects on the lungs, and fish consumption decreases allergic diseases in children and improves the pulmonary function in adults. Therefore, it is important to pay attention to a healthy and appropriate diet to improve the situation in chemical patients.

Factors like excess weight, lack of physical activity, overeating, smoking, fatty foods, lack of fruit and vegetable consumption, dehydration, insufficient energy and protein intake, use of simple sugars, some spices, and fast foods can exacerbate pulmonary and respiratory complications in these patients [70].

Therefore, considering the severity of lesions produced by chemical gases in the body, attention should be paid to the following suggestions to have a healthy diet and decrease respiratory complications:

- It is proved that the use of unsaturated fats has a positive effect on decreasing the inflammation and improving pulmonary diseases. It is recommended that the patients use oil (instead of fat), nuts, and fresh fish.
- Fatty and fried foods exacerbate respiratory symptoms and are not recommended. The use of saturated fats like cream, mayonnaise, and margarine should be minimized or stopped.

- Consumption of large volumes of food increases the gastric volume and decreases the respiratory capacity. Food should be consumed in small, frequent meals.
- Antioxidants and pigments available in spices, like curcumin in turmeric and curry powder, can be used to subside inflammation and enhance the immune system. The use of chili pepper is not recommended due to the stimulation of respiratory problems.
- It is important to enhance the immune system to prevent more severe complications like pulmonary infection. Adequate consumption of fruits and vegetables containing vitamins and antioxidants is recommended (two to four fruits per day or three to five dishes of baked and raw vegetables).
- Vitamins A, C, E, which are potential natural antioxidants, play an important role in decreasing respiratory symptoms. Therefore, the consumption of the sources of vitamin A (yellow, orange, and red fruits vegetables), vitamin C (tomato, cauliflower, grapefruit, kiwifruit, lemon, and citrus fruits), and vitamin E (oil, nuts) on a daily basis is essential [71].
- Weight loss is another approach to alleviate respiratory symptoms in these patients [72]. During the weight loss process, excess body fat is used, respiratory capacity increases, and oxygenation improves. Thus, respiration becomes easier after the loss of excess weight.
- Magnesium helps with airway expansion, vascular dilation, oxygenation, and stress reduction. Magnesium can be found in nuts, dark-green leafy vegetables, and fresh fish.
- Honey, with its anti-inflammatory, antiseptic, and antioxidant properties, can effectively help to reduce respiratory complications. It is recommended to use small amounts of honey every day. Diabetic patients should monitor their blood sugar levels.
- Garlic and lemon have potent anti-inflammatory and antiseptic compounds. They enhance the immune system and dilate airways and thus alleviate respiratory problems. Therefore, their daily consumption is recommended [73].
- Frequent use of cauliflower, garlic, onion, beans, etc. can decrease the respiratory capacity due to abdominal distension. Therefore, they should not be used in large amounts.
- As a result of smoking, toxic particles enter the lungs. These particles have strong oxidative effects; in addition, they stimulate the airways and therefore disturb respiration. Thus, smoking should be strictly avoided.
- Similar to excess weight, malnutrition also has a negative effect on the pulmonary and respiratory function in chemical patients. Adequate food, energy, and protein intake (white meat and low-fat dairy) and consumption of fruits and vegetables are necessary to strengthen the body, enhance the immune system, and prevent infections.
- Stress and anxiety reduction are important ways of decreasing respiratory problems [74]. The use of healthy water, lettuce, borage, and magnesium-containing sources that have tranquilizing effects is strongly recommended.

- Dehydration increases respiratory consequences. It is recommended to drink seven to eight glasses of water per day, and excess use of tea and coffee should be avoided. Drinking water dilutes pulmonary secretions and helps in pulmonary infections.
- The consumption of excess amounts of simple sugars has negative effects on the respiratory trend. Patients should avoid the consumption of sugar, chocolate, cookies, and colas [75].

ANIMAL STUDIES

Many studies have been performed on animals in the areas of prevention and treatment in the acute phase. However, the effectiveness of these methods in human subjects requires further levels of research and controlled clinical trials in exposed patients to enter clinical practice based on evidence-based medicine. One of the most important limitations of these animal studies is the lack of appropriate subjects for long-term evaluation.

In a recent study on laboratory rats, it was observed that the administration of recombinant tissue factor pathway inhibitor improved airway obstruction and gas exchange and decreased mortality in rats exposed to the mustard gas analog [76].

In another study, endotracheal administration of tissue plasminogen activator decreased respiratory distress and improved oxygenation [77]. Vitamin E in guinea pigs improved the tracheal response and regulated cytokines [78]. Moreover, according to a study, the rats that were exposed to a metalloporphyrin catalytic antioxidant after contact with mustard gas had lower levels of oxidative stress indexes and myeloperoxidase in further studies [28].

Table 8.1 summarizes the results of different treatments in some in vitro studies.

FUTURE TREATMENTS

Considering the nature of imbalance between oxidant and antioxidant systems in chemical patients, one of the most important objectives of future treatments is to use drugs that can help to restore the lost balance.

CURCUMIN

There is a great body of evidence that the herbal drug curcumin has anti-inflammatory, antioxidant, and antineoplastic properties through affecting the apoptosis pathway like NF-KB [92,93]. Curcumin affects the TGF-beta/Smads pathway signals and regulates adhering molecules like laminin and cathepsin that are main mediators of mustard injury [94–96]. It has been observed that curcumin improves the response of the airway epithelial cells to the effects of toxins [97] and seems to be an efficient

Table 8.1 The Results of Different Treatments in Some In Vitro Studies

References	Result	Study Sample	Study Design	Author, Year, Country
[79]	Effective	Mice	Dexamethasone liposome-encapsuled vitamin E	Wigenstam et al., 2009, Sweden
[80]	Effective	Rats	Alpha/gamma-tocopherol N-acetyl cysteine + alpha/gamma-tocopherol	Hoesel et al., 2008, USA
[81]	Effective	Chicken embryo	L-thiocitrulline	Sawyer, 1998, Canada
[82]	Effective	Chicken embryo	L-nitroarginine methyl ester	Sawyer et al., 1998, Canada
[29]	Effective	Intracellular macrophages	N-acetylcysteine	Paromov et al., 2008, USA
[83]	Effective	Endothelial cells	N-acetylcysteine	Atkins et al., 2000, USA
[84]	Effective	Alveolar macrophages	Butylated hydroxytoluene N-acetylcysteine	Hultén et al., 1998, Sweden
[85]	Ineffective Effective	Rats	Niacinamide N-acetylcysteine	Anderson et al., 2000, USA
[86]	Effective	Alveolar macrophages	Azithromycin, clarithromycin, erythromycin, roxithromycin	Gao et al., 2010, USA
[87]	Effective	Guinea pigs	Doxycycline	Guignabert et al., 2005, USA
[7]	Effective Mostly effective	Guinea pigs	Surfactant curosurf Salbutamol	Van Helden et al., 2004, Netherlands
[65]	Effective	Human pulmonary epithelial cells	Doxycycline	Raza et al., 2006, USA
[62]	Effective	Rats	Aprotinin	Anderson et al., 2009
[88]	Ineffective	Guinea pigs	Niacinamide	Yourick et al., 1992
[89]	Effective	Rats	Esters of cysteine	Wilde and Upshall, 1991
[90]	Effective	In vitro	Zero-valent iron nanoparticles ferrate(VI)/(III) composite	Zboril et al., 2012
[91]	Effective	Guinea pigs	Nigella sativa Nigella sativa + dexamethasone	Boskabady et al., 2011
[28]	Effective	Rats	Aeolus(AEOL-10150)	O'Neill et al., 2010

option in the treatment of chemical patients. However, its low bioavailability due to hepatic and intestinal metabolism is a major problem [98]. If a correct liposomal form of curcumin is produced, it can be administered orally, intravenously, or through inhalation to be delivered to the target tissue.

HYPERTONIC SALINE

Hypertonic saline solution has been used for septic shock, bronchiolitis, and sputum induction [99]. Its inhalation improves the ciliary function in epithelial cells, which can be helpful [100]. Studies have confirmed the effect of hypertonic saline solution on inflammatory cytokines like IL-8 [101]. Its 5% solution can be safely used in chemical patients as a primary treatment [102]. One of the adverse effects of hypertonic saline solution is hypertension, which should be seriously monitored in these patients.

MANNITOL

Like hypertonic saline solution, inhaled mannitol has different pulmonary effects. It can regulate the flow of the fluids into the air lumen through gradient regulation, dilution of the airway mucus, and improvement of the mucociliary function. Therefore, it can be used for supportive treatment of chronic pulmonary diseases and improvement of airway inflammation. The use of its inhalation formula decreases its possible adverse effects and increases its efficacy [103,104].

REFERENCES

[1] Broumand Ghasemi M, Karamy GH, Pourfarzam SH, Emadi SN, Ghasemi H. Late concurrent ophthalmic, respiratory, coetaneous and psychiatric complications of chemical weapons exposure in 479 war patients. Daneshvar Med 2007;70:81–92.

[2] Kehe K, Szinicz L. Medical aspects of sulfur mustard poisoning. Toxicology 2005;214:198–209.

[3] Abedi AR, Koohestani HR, Roosta Z. The short-term effect of chest physiotherapy on spirometric indices in chemical warfare victims exposed to mustard gas. Armaghanedanesh 2009;3–4:81–91.

[4] Motimer KJ, Harrison TW, Tattersfield AE. Effects of inhaled corticosteroids on bone. Ann Allergy Asthma Immunol 2005;94:15–21.

[5] Balali M, Hefazi M. The pharmacology, toxicology, and medical treatment of sulfur mustard poisoning. Fundam Clin Pharmacol 2005;10:297–315.

[6] Freitag L, Firusian N, Stamatis G, Greschuchna D. The role of bronchoscopy in pulmonary complications due to mustard gas inhalation. Chest 1991;100:1436–41.

[7] Van Helden HP, Kuijpers WC, Diemel RV. Asthma like symptoms following intra tracheal exposure of Guinea pigs to sulfur mustard aerosol: therapeutic efficacy of exogenous lung surfactant curosurf and salbutamol. Inhal Toxicol 2004;16:537–48.

[8] Foroutan A. Iraq chemical war and its medical experiences. Bqiyatallah University Publications; 2003. p. 39–46.
[9] Koch WD. Direkte Kriegserkrankung durch einwirkung chemischer mittel. In: Aschoff L, editor. Pathologische anatomie. Leipzig, Germany: JA Barth; 1921. p. 526–36.
[10] Collumbine H. Medical aspects of mustard gas poisoning. Nature 1947;4031:151–3.
[11] Safarinejad M, Moosavi SA, Montazeri B. Ocular injuries caused by mustard gas: diagnosis, treatment and medical defense. Mil Med 2001;166:67–70.
[12] Borak J, Sidell FR. Agents of chemical warfare: sulfur mustard. Ann Emerg Med 1992;21:303–8.
[13] Wattana M, Bey T. Mustard gas or sulfur mustard: an old chemical agent as a new terrorist treat. Prehosp Disaster Med 2009;24:19–29.
[14] Aghanouri R, Ghanei M, Aslani J, Keivani-Amine H, Rastegar F, Karkhaneh A. Fibrogenic cytokine levels in bronchoalveolar lavage aspirates 15 years after exposure to sulfur mustard. Am J Physiol Lung Cell Mol Physiol 2004;287:1160–4.
[15] Ghanei M, Harandi AA. Long term consequences from exposure to sulfur mustard: a review. Inhal Toxicol May 2007;19(5):451–6.
[16] Poursaleh Z, Harandi AA, Vahedi E, Ghanei M. Treatment for sulfur mustard lung injuries; new therapeutic approaches from acute to chronic phase. Daru September 10, 2012;20(1):27.
[17] Beheshti J, Mark EJ, Akbaei HM, Aslani J, Ghanei M. Mustard lung secrets: long term clinicopathological study following mustard gas exposure. Pathol Res Pract 2006;202(10):739–44.
[18] Ghanei M, Amini Harandi A. The respiratory toxicities of mustard gas. IJMS December 2010;35(4).
[19] D'Urzo AD, Tamari I, Bouchard J, Jhirad R, Jugovic P. New spirometry interpretation algorithm: primary care respiratory alliance of Canada approach. Can Fam Physician October 2011;57(10):1148–52.
[20] Gao X, Ray R, Xiao Y, Barker PE, Ray P. Inhibition of sulfur mustard-induced cytotoxicity and inflammation by the macrolide antibiotic roxithromycin in human respiratory epithelial cells. BMC Cell Biol 2007;8:17.
[21] Yamaryo T, Oishi K, Yoshimine H, Tsuchihashi Y, Matsushima K, Nagatake T. Fourteen-member macrolides promote the phosphatidylserine receptordependentphagocytosis of apoptotic neutrophils by alveolar macrophages. Antimicrob Agents Chemother 2003;47:48–53.
[22] Hodge S, Hodge G, Brozyna S, Jersmann H, Holmes M, Reynolds PN. Azithromycin increases phagocytosis of apoptotic bronchial epithelial cells byalveolar macrophages. Eur Respir J 2006;28:486–95.
[23] Hodge S, Hodge G, Jersmann H, Matthews G, Ahern J, Holmes M, et al. Azithromycin improves macrophage phagocytic function and expression of mannose receptor in chronic obstructive pulmonary disease. Am J Respir Crit Care Med July 15, 2008;178(2):139–48.
[24] Ianaro A, Ialenti A, Maffia P, Sautebin L, Rombolà L, Carnuccio R, et al. Anti-inflammatory activity of macrolide antibiotics. J Pharmacol Exp Ther January 2000;292(1):156–63.
[25] Ghanei M, zadeh MG, Shohrati M. Improvement of respiratory symptoms by long-term low-dose erythromycin in sulfur mustard exposed cases: a pilot study. J Med C B R Def 2005;3:1–9.

[26] Ghanei M, Rajaeinejad M, Motiei-Langroudi R, Alaeddini F, Aslani J. Helium:oxygen versus air:oxygen noninvasive positive-pressure ventilation in patients exposed to sulfur mustard. Heart Lung 2011;40(3):e84–9.

[27] Szabó C, Day BJ, Salzman AL. Evaluation of the relative contribution of nitric oxide and peroxynitrite to the suppression of mitochondrial respirationin immunostimulated macrophages using a manganese mesoporphyrinsuperoxide dismutase mimetic and peroxynitrite scavenger. FEBS Lett 1996;381:82–6.

[28] O'Neill HC, White CW, Veress LA, Hendry-Hofer TB, Loader JE, Min E, et al. Treatment with the catalytic metalloporphyrin AEOL 10150 reduces inflammation and oxidative stress due to inhalation of the sulfur mustard analog 2-chloroethyl ethyl sulfide. Free Radic Biol Med May 1, 2010;48(9):1188–96.

[29] Paromov V, Qui M, Yang H, Smith M, Stone WL. The influence of N-acetyl-L-cysteine on oxidative stress and nitric oxide synthesis in stimulated macrophages treated with a mustard gas analogue. BMC Cell Biol June 20, 2008;9:33.

[30] Grandjean EM, Berthet P, Ruffmann R, Leuenberger P. Efficacy of oral long-term ZV-acetylcysteine in chronicbronchopulmonary disease: a meta-analysis of publisheddouble-blind, placebo-controlled clinical trials. Clin Ther 2000;22:209–21.

[31] Shohrati M, Ghanei M, Shamspour N, Jafari M. Activity and function in lung injuries due to sulphur mustard. Biomarkers November 2008;13(7):728–33.

[32] Jafari M, Ghanei M. Evaluation of plasma, erythrocytes, and bronchoalveolar lavage fluid antioxidant defense system in sulfur mustard-injured patients. Clin Toxicol (Phila) March 2010;48(3):184–92.

[33] Shohrati M, Ghanei M, Shamspour N, Babaei F, Abadi MN, Jafari M, et al. Glutathione and malondialdehyde levels in late pulmonary complications of sulfur mustard intoxication. Lung January–February 2010;188(1):77–83.

[34] Nourani MR, Yazdani S, Roudkenar MH, Ebrahimi M, Halabian R, Mirbagheri L, et al. HO1 mRNA and protein do not change in parallel in bronchial biopsies of patients after long term exposure to sulfur mustard. Gene Regul Syst Bio October 1, 2009;4:83–90.

[35] Mirbagheri L, Habibi Roudkenar M, Imani Fooladi AA, Ghanei M, Nourani MR. Down regulation of super oxide dismutase level in protein might be due to sulfur mustard induced toxicity in lung. Iran J Allergy Asthma Immunol May 15, 2013;12(2):153–60.

[36] Haq F, Mahoney M, Koropatnick J. Signaling events for metallothionein induction. Mutat Res 2003;533(1–2):211–26.

[37] Nourani MR, Ebrahimi M, Roudkenar MH, Vahedi E, Ghanei M, Imani Fooladi AA. Sulfur mustard induces expression of metallothionein-1A in human airway epithelial cells. Int J Gen Med 2011;4:413–9.

[38] Ghanei M, Shohrati M, Jafari M, Ghaderi S, Alaeddini F, Aslani J. N-acetylcysteine improves the clinical conditions of mustard gas-exposed patients with normal pulmonary function test. Basic Clin Pharmacol Toxicol November 2008;103(5):428–32.

[39] Shohrati M, Aslani J, Eshraghi M, Alaedini F, Ghanei M. Therapeutics effect of N-acetyl cysteine on mustard gas exposed patients: evaluating clinical aspect in patients with impaired pulmonary function test. Respir Med March 2008;102(3):443–8.

[40] Ghanei M, Abdolmaali K, Aslani J. Efficacy of concomitant administration of clarythromycin and acetylcysteine in bronchiolitis obliterans in seventeen sulfur mustard-exposed patients: an open-label study. Curr Ther Res 2004;65:495–504.

[41] Urita Y, Watanabe T, Ota H, Iwata M, Sasaki Y, Maeda T, et al. High prevalence of gastroesophageal reflux symptoms in patients with both acute and nonacute cough. Int J Gen Med November 30, 2008;1:59–63.

[42] Ghanei M, Khedmat H, Mardi F, Hosseini A. Distal esophagitis in patients with mustard-gas induced chronic cough. Dis Esophagus 2006;19(4):285–8.
[43] Brackbill RM, Thorpe LE, DiGrande L, Perrin M, Sapp 2nd JH, Wu D, et al. Surveillance for World Trade Center disaster health effects among survivors of collapsed and damaged buildings. MMWR Surveill Summ April 7, 2006;55(2):1–18.
[44] Ghanei M, Harandi AA, Tazelaar HD. Isolated bronchiolitis obliterans: high incidence and diagnosis following terrorist attacks. Inhal Toxicol April 2012;24(5):340–1.
[45] Aliannejad R, Hashemi-Bajgani SM, Karbasi A, Jafari M, Aslani J, Salehi M, et al. GERD related micro-aspiration in chronic mustard-induced pulmonary disorder. J Res Med Sci August 2012;17(8):777–81.
[46] Galmiche JP, Zerbib F, Varannes S. Review article: respiratory manifestations of gastro-oesophageal reflux disease. Aliment Pharmacol Ther 2008;27:449–64.
[47] Theodoropoulos DS, Ledford DK, Lockey RF, Pecoraro DL, Rodriguez JA, Johnson MC, et al. Prevalence of upper respiratory symptoms in patients with symptomatic gastroesophageal reflux disease. Am J Respir Crit Care Med July 1, 2001;164(1):72–6.
[48] Niewoehner DE, Erbland ML, Deupree RH, Collins D, Gross NJ, Light RW, et al. Effect of systemic glucocorticoids on exacerbations of chronic obstructive pulmonary disease. Department of Veterans Affairs Cooperative Study Group. N Engl J Med June 24, 1999;340(25):1941–7.
[49] Emerman CL, Connors AF, Lukens TW, May ME, Effron D. A randomized controlled trial of methylprednisolone in the emergency treatment of acute exacerbations of COPD. Chest March 1989;95(3):563–7.
[50] Ghanei M, Khalili AR, Arab MJ, Mojtahedzadeh M, Aslani J, Lessan-Pezeshki M, et al. Diagnostic and therapeutic value of short-term corticosteroid therapy in exacerbation of mustard gas-induced chronic bronchitis. Basic Clin Pharmacol Toxicol November 2005;97(5):302–5.
[51] Calverley P, Pauwels R, Vestbo J, Jones P, Pride N, Gulsvik A, et al. TRial of Inhaled STeroids ANd long-acting beta2 agonists study group. Combined salmeterol and fluticasone in the treatment of chronic obstructive pulmonary disease: a randomised controlled trial. Lancet February 8, 2003;361(9356):449–56.
[52] Ghanei M, Tazelaar HD, Chilosi M, Harandi AA, Peyman M, Akbari HM, et al. An international collaborative pathologic study of surgical lung biopsies from mustard gas-exposed patients. Respir Med June 2008;102(6):825–30.
[53] Wang L, Hollenbeak CS, Mauger DT, Zeiger RS, Paul IM, Sorkness CA, et al. Childhood Asthma Research and Education Network of the National Heart, Lung, and Blood Institute. Cost-effectiveness analysis of fluticasone versus montelukast in children with mild-to-moderate persistent asthma in the Pediatric Asthma Controller Trial. J Allergy Clin Immunol January 2011;127(1):161–6. 166.e1.
[54] Ghanei M, Shohrati M, Harandi AA, Eshraghi M, Aslani J, Alaeddini F, et al. Inhaled corticosteroids and long-acting beta 2-agonists in treatment of patients with chronic bronchiolitis following exposure to sulfur mustard. Inhal Toxicol August 2007;19(10):889–94.
[55] Sohrabpour H, Zamir Roshan F, Aminorroaya AP. Comparison of acute bronchodilatory effects of inhaled salbotamol and combivent in mustard gas victims. Iran J Med Sci 1996;21:29–34.
[56] Farncombe M, Chater S, Gillin A. The use of nebulized opioids for breathlessness: a chart review. Palliat Med 1994;8(4):306–12.
[57] Bruera E, Macmillan K, Pither J, MacDonald RN. Effects of morphine on the dyspnea of terminal cancer patients. J Pain Symptom Manag 1990;5(6):341–4.

[58] Shohrati M, Ghanei M, Harandi AA, Foroghi S, Harandi AA. Effect of nebulized morphine on dyspnea of mustard gas-exposed patients: a double-blind randomized clinical trial study. Pulm Med 2012;2012:610921.
[59] Ghanei M, Panahi Y, Mojtahedzadeh M, Khalili AR, Aslani J. Effect of gamma interferon on lung function of mustard gas exposed patients, after 15 years. Pulm Pharmacol Ther 2006;19(2):148–53.
[60] Ghanei M, Panahi Y, Aslani J, Mojtahedzadeh M. Successful treatment of pulmonary obstructive lesion in chemical warfare casualties with gamma-interferon. Kowsar Med J 2003;2:151–7.
[61] Cowan FM, Anderson DR, Broomfield CA, Byers SL, Smith WJ. Biochemical alterations in rat lung lavage fluid following acute sulfur mustard inhalation:II. Increase in proteolytic activity. Inhal Toxicol 1997;9:53–61.
[62] Anderson DR, Taylor SL, Fetterer DP, Holmes WW. Evaluation of protease inhibitors and an antioxidant for treatment of sulfur mustard-induced toxic lung injury. Toxicology 2009;263:41e6.
[63] Amin AR, Attur MG, Thakker GD, Patel PD, Vyas PR, Patel RN, et al. A novel mechanism of action of tetracyclines: effects on nitric oxide synthases. Proc Natl Acad Sci USA November 26, 1996;93(24):14014–9.
[64] Weinberger B, Laskin JD, Sunil VR, Sinko PJ, Heck DE, Laskin DL. Sulfur mustard-induced pulmonary injury: therapeutic approaches to mitigating toxicity. Pulm Pharmacol Ther 2011;24(1):92–9.
[65] Raza M, Ballering JG, Hayden JM, Robbins RA, Hoyt JC. Doxycycline decreases monocyte chemoattractant protein-1 in human lung epithelial cells. Exp Lung Res 2006;32:15e26.
[66] Hassan ZM, Ebtekar M, Ghanei M, Taghikhani M, Noori Daloii MR, Ghazanfari T. Immunobiological consequences of sulfur mustard contamination. Iran J Allergy Asthma Immunol September 2006;5(3):101–8.
[67] Brockmeyer M. Immunomodulatory properties of Cimetidine in ARC patients. Clin Immunol Immunother 1988;48:50–60.
[68] Fallah Mohammadi Z, Tari M. Effect of a cycle of selective aerobic treadmill exercise on some PFT indexes in chemical patients (mustard gas). Sports Physiol 2009;22(6):13–27.
[69] Khosravi M, Tavakkoli M. Physical exercise considerations in chemical patients. In: Scientific research Seminar of Medicine and Sacrifice. December 2012. Mashhad, Iran. 2012.
[70] Romieu I, Trenga C. Diet and obstructive lung diseases. Epidemiol Rev 2001;23(2):268–87.
[71] Keranis E, Makris D, Rodopoulou P, Martinou H, Papamakarios G, Daniil Z, et al. Impact of dietary shift to higher-antioxidant foods in COPD: a randomised trial. Eur Respir J October 2010;36(4):774–80.
[72] Lari SM, Ghobadi H, Attaran D, Kazemzadeh A, Mahmoodpour A, Shadkam O, et al. The significance of BODE (BMI, obstruction, dyspnea, exercise) index in patients with mustard. Lung 2013;1(1):7–11.
[73] George J, Ioannides-Demos LL, Santamaria NM, Kong DC, Stewart K. Use of complementary and alternative medicines by patients with chronic obstructive pulmonary disease. Med J Aust September 6, 2004;181(5):248–51.
[74] Laurin C, Moullec G, Bacon SL, Lavoie KL. Impact of anxiety and depression on chronic obstructive pulmonary disease exacerbation risk. Am J Respir Crit Care Med May 1, 2012;185(9):918–23.

[75] Namavar S. http://www.fashnews.ir/archive/health/13887-2012-05-20-08-31-25.html; May 2012.

[76] Rancourt RC, Veress LA, Ahmad A, Hendry-Hofer TB, Rioux JS, Garlick RB, et al. Tissue factor pathway inhibitor prevents airway obstruction, respiratory failure and death due to sulfur mustard analog inhalation. Toxicol Appl Harmacol October 1, 2013;272(1):86–95.

[77] Veress LA, Hendry-Hofer TB, Loader JE, Rioux JS, Garlick RB, White CW. Tissue plasminogen activator prevents mortality from sulfur mustard analog–induced airway obstruction. Am J Respir Cell Mol Biol April 2013;48(4):439–47.

[78] Boskabady MH, Amery S, Vahedi N, Khakzad MR. The effect of vitamin E on tracheal responsiveness and lung inflammation in sulfur mustard exposed guinea pigs. Inhal Toxicol February 2011;23(3):157–65.

[79] Wigenstam E, Rocksén D, Ekstrand-Hammarström B, Bucht A. Treatment with dexamethasone or liposome-encapsuled vitamin E provides beneficial effects after chemical-induced lung injury. Inhal Toxicol 2009;21(11):958–64.

[80] Hoesel LM, Flierl MA, Niederbichler AD, Rittirsch D, McClintock SD, Reuben JS, et al. Ability of antioxidant liposomes to prevent acute and progressive pulmonary injury. Antioxid Redox Signal 2008;10(5):973–81.

[81] Sawyer TW. Characterization of the protective effects of L-nitroarginine methyl ester (L-NAME) against the toxicity of sulphur mustard in vitro. Toxicology 1998;13(1):21–32.

[82] Sawyer TW, Hancock JR, D'Agostino PA. l-thiocitrulline. A potent protective agent against the toxicity of sulphur mustard in vitro. Toxicol Appl Pharmacol 1998;151(2):340–6.

[83] Atkins KB, Lodhi IJ, Hurley LL, Hinshaw DB. N-Acetylcysteine and endothelial cell injury by sulfur mustard. J Appl Toxicol 2000;20(1):125–8.

[84] Hultén LM, Lindmark H, Scherstén H, Wiklund O, Nilsson FN, Riise GC. Butylated hydroxytoluene and N-acetylcysteine attenuates tumor necrosis factor-alpha (TNF-alpha) secretion and TNF-alpha mRNA expression in alveolar macrophages from human lung transplant recipients in vitro. Transplantation 1998;66(3):364–9. 15.

[85] Anderson DR, Byers SL, Vesely KR. Treatment of sulfur mustard (HD)-induced lung injury. J Appl Toxicol 2000;20(Suppl. 1):S129–32.

[86] Gao X, Ray R, Xiao Y, Ishida K, Ray P. Macrolide antibiotics improve chemotactic and phagocytic capacity as well as reduce inflammation in sulfur mustard-exposed monocytes. Pulm Pharmacol Ther 2010;23(2):97–106.

[87] Guignabert C, Taysee L, Calvet JH, Planus E, Delamanche S, Galiacy S, et al. Effect of doxycycline on sulfur mustard-induced respiratory lesions in guinea pigs. Am J Physiol Cell Mol Physiol 2005;289:L67–74.

[88] Yourick JJ, Dawson JS, Mitcheltree LW. Sulfur mustard-induced microvesication in hairless guinea pigs: effect of short-term niacinamide administration. Toxicol Appl Pharmacol 1992;117(1):104–9.

[89] Wilde PE, Upshall DG. Cysteine esters protect cultured rodent lung slices from sulphur mustard. Hum Exp Toxicol 1994;13(11):743–8.

[90] Zboril R, Andrle M, Oplustil F, Machala L, Tucek J, Filip J, et al. Treatment of chemical warfare agents by zero-valent iron nanoparticles and ferrate(VI)/(III) composite. J Hazard Mater 2012;211–212:126–30.

[91] Boskabady MH, Vahedi N, Amery S, Khakzad MR. The effect of Nigella sativa alone, and in combination with dexamethasone, on tracheal muscle responsiveness and lung inflammation in sulfur mustard exposed guinea pigs. J Ethnopharmacol 2011;37(2):1028–34.

[92] Das L, Vinayak M. Anti-carcinogenic action of curcumin by activation of antioxidant defence system and inhibition of NF-kappaB signalling in lymphoma-bearing mice. Biosci Rep 2012;32(2):161–70.
[93] Schaffer M, Schaffer PM, Zidan J, Bar Sela G. Curcuma as a functional food in the control of cancer and inflammation. Curr Opin Clin Nutr Metab Care 2011;14(6):588–97.
[94] Li Y, Chen ZQ, Li YD. Effects of curcumin on the epithelial mesenchymal transition and TGF-beta/Smads signaling pathway in unilateral ureteral obstruction rats. Zhongguo Zhong Xi Yi Jie He Za Zhi 2011;31(9):1224–8.
[95] Adelipour M, Imani Fooladi AA, Yazdani S, Vahedi E, Ghanei M, Nourani MR. Smad molecules expression pattern in human bronchial airway induced by sulfur mustard. Iran J Allergy Asthma Immunol 2011;10(3):147–54.
[96] Zhang D, Huang C, Yang C, Liu RJ, Wang J, Niu J, et al. Antifibrotic effects of curcumin are associated with overexpression of cathepsins K and L in bleomycin treated mice and human fibroblasts. Respir Res November 29, 2011;12:154.
[97] Rennolds J, Malireddy S, Hassan F, Tridandapani S, Parinandi N, Boyaka PN, et al. Curcumin regulates airway epithelial cell cytokine responses to the pollutant cadmium. Biochem Biophys Res Commun January 6, 2012;417(1):256–61.
[98] Sharma RA, Steward WP, Gescher AJ. Pharmacokinetics and pharmacodynamics of curcumin. Adv Exp Med Biol Rev 2007;595:453–70.
[99] Hom J, Fernandes RM. When should nebulized hypertonic saline solution be used in the treatment of bronchiolitis? Paediatr Child Health 2011;16(3):157–8.
[100] Burrows EF, Southern KW, Noone PG. Sodium channel blockers for cystic fibrosis. Cochrane Database Syst Rev March 14, 2012;3:CD005087.
[101] Reeves EP, Williamson M, O'Neill SJ, Greally P, McElvaney NG. Nebulized hypertonic saline decreases IL-8 in sputum of patients with cystic fibrosis. Am J Respir Crit Care Med 2011;183(11):1517–23.
[102] Al-Ansari K, Sakran M, Davidson BL, El Sayyed R, Mahjoub H, Ibrahim K. Nebulized 5% or 3% hypertonic or 0.9% saline for treating acute bronchiolitis in infants. J Pediatr 2010;157(4):630–4.
[103] Hurt K, Bilton D. Inhaled mannitol for the treatment of cystic fibrosis. Expert Rev Respir Med 2012;6(1):19–26.
[104] de Nijs SB, Fens N, Lutter R, Dijkers E, Krouwels FH, Smids-Dierdorp BS, et al. Airway inflammation and mannitol challenge test in COPD. Respir Res January 18, 2011;12:11.

CHAPTER

9

Mustard Gas and Cancer in Chemical Patients

The hypothesis of sulfur mustard (SM) alkalization was first proposed in 1968. The hypothesis states that with alkalization of guanine in the O_6 site, it is no longer possible to repair it with O_2-alkylguanine-DNA alkyltransferase and as a result, the grounds are prepared for SM carcinogenesis. Fig. 3.3 shows the reaction of mustard gas with the DNA structure.

Moreover, defects in the DNA repair after exposure to mustard gas result in cellular death through apoptosis. The removal of the alkalized cells through apoptosis on the one hand and restoration of the possibility of DNA repair in the course of time after exposure on the other hand, in case of one time of exposure, are among the reasons for the low carcinogenesis potential of SM.

Many important cellular-molecular studies on the effects of mustard gas on the body, its complications, and treatments have been and are still being performed in Iran. Another investigated hypothesis is the induction of mutations in human tumor suppressors and oncogenes. Among them, P_{53} is a protein with a tumor-suppressor role. When a cell is exposed to stress, P_{53} accumulates in the cell due to inhibition of transcription. In normal conditions, this protein regulates the events leading to apoptosis and cell cycle arrest. It is obvious that cells that are deficient in P_{53} are genetically unstable and more prone to cancer.

In fact, the mechanism of the carcinogenesis of mustard gas begins with cyclization of SM in the aqueous environment of a victim to a highly reactive episulfonium ion that might alkylate DNA. If the repair process does not correct the defect, changes occur in the nucleoside unit with a "G" to "A" change being the most common [1].

In a study on tumor suppressors in Iran, 20 chemical patients exposed to SM that had lung cancer were evaluated. The specimens obtained from these individuals were investigated for mutation in P_{53} and KRAS. Of 16 acceptable specimens, five had mutations in P_{53} with a G to A change in most of them [2] while no mutations were detected in KRAS. P_{53} mutation was also reported in the Japanese workers of a SM production factory in 1994. Furthermore, another study showed that cigarette smoking increased P_{53} mutation in patients with a history of exposure to mustard gas.

There were concerns about the carcinogenesis of mustard gas from the early years due to subacute and chronic consequences of this substance.

Mustard Lung. http://dx.doi.org/10.1016/B978-0-12-803952-6.00009-5

We will evaluate the effects of mustard gas in the following groups:

1. Workers in mustard gas–producing factories who were contaminated with mustard gas due to inappropriate protective clothing.
2. Soldiers and residents of war zones that were exposed to high-dose mustard gas in a short time.

It should be noted that the main objective of this chapter is to evaluate lung cancer resulting from mustard gas.

OCCUPATIONAL EXPOSURE

The main concerns regarding the carcinogenesis of mustard gas are about the reported cases of cancer in the workers of a mustard-producing factory. These workers were exposed to low levels of mustard gas for a long time. Most of the studies were performed on Japanese, British, and German workers. The results showed a high percentage of different cancers in different organs, especially lung cancer.

The Japanese army established a chemical gas factory in 1929–1945. Many factory workers were exposed to SM for 7–9 years due to inappropriate clothing. A considerable increase in lung cancer was reported in these workers 17–20 years later [3,4].

A considerable percentage of the British workers who worked in mustard gas–producing factories for 4–5 years developed lung cancer, but its prevalence was lower when compared with Japanese workers, which could be due to improved health and hygiene conditions and better protection.

In another study on 502 British workers during 5 years from 1940 to 1945, the number of patients with tracheal and laryngeal cancer was found to be high. In another study on 3345 British workers, 11, 15, and 200 deaths were reported due to laryngeal, pharyngeal, and pulmonary cancer, respectively. It should be noted that these studies only evaluated the mortality of the involvement of the respiratory system; investigation of the involvement of other organs was not the objective of these studies [5].

In another study that was performed on 245 German workers over 20 years there was a statistically significant increase in malignant tumors, especially bronchial carcinoma, bladder carcinoma, and leukemia [6]. Other studies in this regard also reported a high incidence of cancer in individuals exposed to low-dose mustard gas for a long time, mainly because these people do not have the necessary time for DNA repair due to repeated exposures.

The high incidence of cancer upon long time exposure indicated the adverse and hazardous effects of mustard gas on the personnel and its devastating power in the battlefield.

EXPOSURE TO HIGH-DOSE MUSTARD GAS

This section is about soldiers and civilians exposed to mustard gas during war. The exposure dose is sometimes so high that it results in the victim's death immediately due to respiratory distress and bronchospasm. One of its important and obvious symptoms is severe skin blisters like in burns.

When the reports of the high prevalence of cancer in workers of mustard-producing factories were published, concerns were raised that soldiers that were exposed to low- or high-dose mustard gas only once might be at risk of different cancers, including lung cancer. Different studies were conducted to investigate this possibility. A number of important studies in this area are presented in the following discussion.

A well-designed study compared US soldiers exposed to mustard gas with a similar group of soldiers with a negative history of exposure but who developed influenza and the resulting pneumonia in the influenza epidemic in 1918 [7]. The frequency of lung cancer was 1.47 and 0.8 in the exposed and nonexposed groups, respectively. A 10-year follow-up also showed similar results. According to the experts, the results of the study were not statistically significant and no death was reported due to cancer. This study had some limitations and factors like smoking, alcohol consumption, and the history of exposure to other pollutants were not evaluated. Death due to lung cancer was observed in 25% of the soldiers exposed to mustard gas, 18% of the soldiers with a history of pneumonia, and 19% of the controls. The risk of death of lung cancer was estimated at one-third among exposed soldiers compared with controls. Mustard gas did not show strong carcinogenesis effects in this study in comparison with a study performed by Lea and Case in 1995. In this study, 1267 British men who participated in the war between 1914 and 1918 were evaluated, and it was concluded that the mortality from lung cancer was high both in exposed soldiers and in patients with bronchitis who had no exposure, with no significant difference between the two groups. The authors stated that pulmonary tissue injury due to exposure to mustard gas or chronic bronchitis could cause lung cancer. They also hypothesized that although the dose of mustard was not high, it was enough for carcinogenesis [8].

However, with underlying pulmonary diseases like chronic bronchitis due to smoking, mustard exposure can be carcinogenic, although it is not true for all pulmonary diseases. For example, in cystic fibrosis, although there is tissue injury, there is no risk of pulmonary cancer because of the lack of a carcinogenic factor.

In a primary study conducted in 1975, the relative risk of lung cancer in two groups of smokers and nonsmokers that were both exposed to mustard gas was evaluated. In fact, the synergistic effect of smoking and mustard gas on carcinogenesis was investigated in this study. The results showed that the relative risk was 4.3 in the first and 4.4 in the second group with no significant difference between them. According to this study, coexistence of smoking and mustard exposure did not increase the risk of lung cancer when compared to their effects individually. In a new study based on P_{53} activity, the primary evidence indicated that smoking could increase the carcinogenic risk of mustard gas [2].

Many years ago when the devastating effects of mustard gas were not yet identified, in a clinical trial study in the United States, about 60,000 individuals were directly or indirectly exposed to mustard gas to evaluate its effects and identify the best protective clothing. Unfortunately, the study showed that mustard gas penetrated even the protective clothes of the volunteers. In a study that was performed in 2000 to evaluate their health after 50 years, Bullman and Kang evaluated 1545 US Navy SEALs who were exposed to low-dose mustard gas during these experiments in

World War II. These soldiers were compared with 2663 individuals with a negative history of exposure. The results showed no difference in the prevalence of death due to cancer between the two groups [9]. However, focus on death and not carcinogenesis, and lack of attention to the history of smoking, alcohol consumption, and exposure and contact to other carcinogens were the weak points of this study.

The biggest tragedy occurred during the Iraq–Iran war of 1980–1988. The Iraqi army used mustard gas against Iranian soldiers and even innocent civilians many times. As a result, about 100,000 people were exposed to mustard gas, which resulted in the deaths of many victims at the time of exposure and injury or disability in about 50,000 people. For this reason, many studies were performed on the nature of mustard toxicity and its complications and effective treatments in Iran.

In one study, 197 chemical patients of the Iraq–Iran war were compared with 86 healthy controls. The results showed no evidence of pulmonary or bronchial cancer 10 years after exposure to mustard gas [10]. In another cross-sectional study, 98 patients with severe hemoptysis who had a history of exposure to mustard gas were investigated meticulously for evidence of cancer. The bronchial lavage fluid was negative for cancer cells in all patients. Imaging was also negative for malignancy. On pathological study, the report of 9%, 83%, and 8% of the biopsy specimens was normal, chronic inflammation, and squamous metaplasia, respectively. It was concluded that severe hemoptysis was not an appropriate index for lung cancer although it is necessary to follow up these patients for malignancies [11].

In a historical cohort study, 500 men who were exposed to high-dose mustard gas in the war were compared with a similar control group in terms of demographic characteristics [12]. Despite an interval of 18 years between the exposure and study, only three cases of cancer, two cases of lung cancer, and one case of lymphoma, were detected. The relative risk of cancer was 4.02, which was not statistically significant. In another study performed on 500 chemical patients including 372 civilians living in Sardasht, a city near the Iran–Iraq border, no case of cancer was found [13].

There was no evidence of mustard carcinogenesis following one acute exposure until a few years ago [14]. However, new studies have produced different results suggesting the possibility of carcinogenesis years after the exposure.

In a study performed on chemical patients of the Iraq–Iran war exposed to a single dose of mustard gas who suffered from respiratory problems, P_{53} immunoreactivity was evaluated in the bronchial epithelium of people with a history of smoking or exposure to mustard gas. In this study, 68 chemical patients were divided into two groups: 35 patients with a history of exposure (including eight smokers) and 33 patients with no exposure (16 smokers). Pulmonary function tests were used to investigate the severity of the disease. The P_{53} protein obtained from the bronchial tissue was measured using immunohistochemistry staining. Among nonsmokers, P_{53} gene expression was observed in 41.2% of patients with a negative history of exposure and 14.8% of the exposed patients. Among smokers, it was seen in 25% of nonexposed and 50% of the exposed patients. Initial data trends suggested an additive contribution of SM exposure and smoking to P_{53} immunoreactivity [2].

In another study, the demographic data and tumor specimens of 20 Iranian male patients with lung cancer with only exposure to high-dose mustard gas during the

Iraq–Iran war were collected. Biopsy specimens of lung cancer were evaluated for P_{53} protein using immunohistochemistry. Moreover, the extracted DNA from the tissue was studied for mutation in the P_{53} and KRAS gene using polymerase chain reaction. The relative young age of developing lung cancer (range: 27–72 years, mean age: 48 years) in mustard victims, especially in nonsmokers (mean age: 40.7 years), may indicate a specific cause. The disease was developed before the age of 40 years in 7 out of 20 patients. Five out of 16 patients with cancer whose data of DNA sequencing was extracted had evidence of P_{53} mutation (in exons 5–8) due to exposure to mustard gas. Two patients with lung cancer had several point mutations in P_{53}, similar to the results of the workers of a SM production factory. No mutations were detected in the KRAS gene. It was concluded that distinct features of lung cancer in these patients indicated that even exposure to a single dose of mustard gas might increase the risk of lung cancer in some victims [15].

We conducted a cohort study to compare the incidence rates of malignant disorders in 7570 veterans exposed to SM and 7595 unexposed comrades in a 25-year follow-up period. We also determined the hazard ratio of cancer occurrence for SM exposure during the follow-up period.

The results showed that the prevalence of cancer increased markedly following exposure to mustard gas. The incidence rate ratio of cancer for SM exposure was 1.81 (95% CI 1.27–2.56), and the age-adjusted incidence rate ratio was 1.64 (95% CI 1.15–2.34).The hazard ratio of cancer was 2.02 (95% CI 1.41–2.88) [16]. The role of time and confounding factors like the presence of inflammation and chronic diseases in the occurrence of cancer was not evaluated in our study.

CONCLUSIONS

The risk of the carcinogenesis from SM in the human respiratory system following frequent contact even with low doses of this chemical compound, like the workers of the SM–producing factory, is proved in different studies. On the other hand, according to the available evidence, the possibility of mustard gas carcinogenesis following one exposure even to a high-dose of mustard gas has not been proved, and it does not increase the risk of pulmonary cancer according to the published data. However, considering the new reports indicating the development of cancer in the victims of the Iraq–Iran war, the possibility of carcinogenesis has increased in these patients. Cohort studies are required to make a definite statement on the carcinogenesis of mustard gas following one exposure [14], and the patients should be carefully screened for possible malignancies periodically.

CANCER IN OTHER ORGANS

Late complications of mustard gas include its effects on the respiratory and digestive systems, skin, and the incidence of different cancers. Mustard gas is definitely a very strong mutagen and its mutagenic effects are proved in in vitro studies on bacteria, human cells, *Drosophila*, and plants. There are diverse reports on the development of cancer following acute short-term exposure to mustard gas. Many studies have reported that cancer patients treated with mustard family compounds (nitrogen mustard) have developed secondary malignancies, mainly blood cancers and lymphoma, in the long term.

In a study by Balali-Mood, the incidence of acute myeloblastic leukemia and acute lymphoblastic leukemia was 0.23% and 2%, which was 18 and 12 times more when compared to the control group, respectively. On the other hand, a 300-times increase was observed in the risk of the incidence of chronic myeloblastic lymphoma when compared to the control group [17]. However, no skin malignancies were detected when evaluating the cutaneous lesions of the chemical patients 10 years after exposure to mustard gas. In this study, only three cancer patients were detected in the exposure group, one case of lymphoma and two cases of lung cancer. All three patients had died. In general, 10 patients in the exposure group and seven patients in the nonexposure group had died.

Considering different studies, the incidence of cancer according to person-years was 33.85 in 100,000 in the exposure group and the mortality rate in the exposure group was 112.84 per 100,000 population. However, comparison of the cancer incidence in the exposed patients with national statistics, after adjustment for age, does not show any significant relationship between acute exposure to SM, even at high doses, and the incidence of malignancy.

Evaluation of chemical patients for the occurrence of cutaneous cancers also suggests that the risk of skin carcinoma after one acute exposure does not seem to increase in these patients.

REFERENCES

[1] Shibata MA, Shirai T, Ogawa K, Takahashi S, Wild CP, Montesano R, et al. DNA methylation adduct formation and H-ras gene mutations in progression of N-butyl-N-(4-hydroxybutyl) nitrosamine-induced bladder tumors caused by a single exposure to N-methyl-N-nitrosourea. Carcinogenesis December 1994;15(12):2965–8.

[2] Hosseini-Khalili A, Haines DD, Modirian E, Soroush M, Khateri S, Joshi R, et al. Mustard gas exposure and carcinogenesis of lung. Mutat Res August 1, 2009;678(1):1–6.

[3] Nishimoto Y, Yamakiw M, Shigenobu T, Yuketake M, Matsusaka S. Cancer of the respiratory tract observed in workers having returned from a poison gas factory. Gun Kaguku Ryoho 1986;13:1144–8.

[4] Wada S, Nishimoto Y, Miyanshi M, Katsuta S, Nishiki M, Yamada A, et al. Malignant respiratory tract neoplasms related to poison gas exposure. Hiroshima J Med Sci 1962;11:81–91.

[5] Easton DF, Peto J, Doll R. Cancers of the respiratory tract in mustard gas workers. Br J Ind Med October 1988;45(10):652–9.

[6] Weiss A, Weiss B. Carcinogenesis due to mustard gas exposure in man, important sign for therapy with alkylating agents. Dtsch Med Wochenschr April 25, 1975;100(17):919–23.

[7] Beebe GW. Lung cancer in world war I veterans: possible relation to mustard-gas injury and 1918 influenza epidemic. J Natl Cancer Inst December 1960;25:1231–52.

[8] Case RA, Lea AJ. Mustard gas poisoning, chronic bronchitis, and lung cancer; an investigation into the possibility that poisoning by mustard gas in the 1914–18 war might be a factor in the production of neoplasia. Br J Prev Soc Med April 1955;9(2):62–72.

[9] Bullman T, Kang H. A fifty year mortality follow-up study of veterans exposed to low level chemical warfare agent, mustard gas. Ann Epidemiol July 2000;10(5):333–8.

[10] Emad A, Rezaian GR. The diversity of the effects of sulfur mustard gas inhalation on respiratory system 10 years after a single, heavy exposure: analysis of 197 cases. Chest 1997;112(3):734–8.

[11] Ghanei M, Eshraghi M, Jalali AR, Aslani J. Evaluation of latent hemoptysis in sulfur mustard injured patients. Environ Toxicol Pharmacol 2006;22(2):128–30.

[12] Gilasi HR, Holakouie Naieni K, Zafarghandi MR, Mahmoudi M, Ghanei M, Soroush MR, et al. Relationship between mustard gas and cancer in Iranian soldiers of imposed war in Isfahan province: a pilot study. J Sch Public Health Inst Public Health Res 2006;4(3):15–24.

[13] Ghazanfari T, Faghihzadeh S, Aragizadeh H, Soroush MR, Yaraee R, Mohammad Hassan Z, et al. Sardasht-Iran cohort study research group. Sardasht-Iran cohort study of chemical warfare victims: design and methods. Arch Iran Med January 2009;12(1):5–14.

[14] Ghanei M, Harandi AA. Lung carcinogenicity of sulfur mustard. Clin Lung Cancer January 2010;11(1):13–7.

[15] Ghanei M, Amiri S, Akbari H, Kosari F, Khalili AR, Alaeddini F, et al. Correlation of sulfur mustard exposure and tobacco use with expression (immunoreactivity) of p53 protein in bronchial epithelium of Iranian "mustard lung" patients. Mil Med January 2007;172(1):70–4.

[16] Zafarghandi MR, Soroush MR, Mahmoodi M, Naieni KH, Ardalan A, Dolatyari A, et al. Incidence of cancer in Iranian sulfur mustard exposed veterans: a long-term follow-up cohort study. Cancer Causes Control January 2013;24(1):99–105.

[17] Balali-Mood M. First report of delayed toxic effects of yperite poisoning in Iranian fighters. In: Heyndricks B, editor. Terrorism: analysis and detection of explosives. Proceedings of the second world congress on new compounds in biological and chemical warfare. Gent: Rijksuniversiteit; 1986. p. 489–95.

Appendix

PRESCRIPTION FOR DYSPNEA EXACERBATION AND INCREASED SPUTUM PRODUCTION

1. Tab Prednisolone 50 mg N = 10

   1 tablet per day for 5 days.

2. Spray Fluticasone N = 2

   3 puffs every 12 h.

3. Spray Salmeterol N = 1

   2 puffs every 12 h.

4. Cap Azithromycin N = 60

   1 capsule per day.

5. Tab N-Acetylcysteine 600 mg N = 100

   1 tablet dissolved in a glass of water every 12 h.

PRESCRIPTION FOR DYSPNEA EXACERBATION

1. Tab Prednisolone Fort 50 mg N = 10

   1 tablet per day for 5 days.

2. Spray Fluticasone N = 2

   3 puffs every 12 h.

3. Spray Salmeterol N = 1

   2 puffs every 12 h.

4. Tab N-Acetylcysteine 600 mg N = 100

   1 tablet dissolved in a glass of water every 12 h.

PRESCRIPTION FOR DYSPNEA EXACERBATION CONCOMITANT WITH REFLUX EXACERBATION

1. Tab Prednisolone Fort 50 mg N = 10

   1 tablet per day for 5 days.

2. Spray Fluticasone N = 2

   3 puffs every 12 h.

3. Spray Salmeterol N = 1

   2 puffs every 12 h.

4. Cap Omeprazole 20 mg N = 60

   1 tablet in the morning and 1 tablet at night.

5. Tab Domperidone 10 mg N = 30

   1 tablet at night.

6. Tab N-Acetylcysteine 600 mg N = 100

   1 tablet dissolved in a glass of water every 12 h.

Index

'*Note*: Page numbers followed by "f" indicate figures and "t" indicate tables.'

N

O

P

R

S

T

V

Printed in the United States
By Bookmasters